ATLAS OF SKELETAL MUSCLES

THIRD EDITION

ROBERT J. STONE
Suffolk County Community College

JUDITH A. STONE
Suffolk County Community College

McGraw Hill

Boston Burr Ridge, IL Dubuque, IA Madison, WI New York San Francisco St. Louis
Bangkok Bogotá Caracas Lisbon London Madrid
Mexico City Milan New Delhi Seoul Singapore Sydney Taipei Toronto

McGraw-Hill Higher Education
A Division of The ***McGraw-Hill*** *Companies*

ATLAS OF SKELETAL MUSCLES, THIRD EDITION

This book is printed on recycled, acid-free paper containing 10% postconsumer waste.

3 4 5 6 7 8 9 0 QPD/QPD 0 9 8 7 6 5 4 3 2 1

ISBN 0–07–290332–5

Vice president and editorial director: *Kevin T. Kane*
Publisher: *Colin H. Wheatley*
Sponsoring editor: *Kristine Tibbetts*
Developmental editor: *Patrick F. Anglin*
Marketing manager: *Heather K. Wagner*
Editing associate: *Joyce Watters*
Senior production supervisor: *Mary E. Haas*
Coordinator of freelance design: *Rick Noel*
Compositor: *Shepherd, Inc.*
Typeface: *10/12 Geneva II Light*
Printer: *Quebecor Printing Book Group/Dubuque, IA*

Interior design: *Kathy Theis*
Cover design: *Sean M. Sullivan*

Library of Congress Cataloging-in-Publication Data

Stone, Robert J.
Atlas of skeletal muscles / Robert J. Stone, Judith A. Stone.—
3rd ed.
p. cm.
Includes index.
ISBN 0–07–290332–5
1. Striated muscle Atlases. I. Stone, Judith A.
[DNLM: 1. Musculoskeletal System Atlases. WE 17 S879a 2000]
QM151.S76 2000
611'.73'0222—dc21
DNLM/DLC
for Library of Congress 99–27910
CIP

INTERNATIONAL EDITION ISBN 0–07–116992–X

www.mhhe.com

DEDICATION

To Karen, Andrew, and Laura for their interest, enthusiasm, and cooperation.

CONTENTS

CHAPTER FOUR MUSCLES OF THE NECK 49

CHAPTER FIVE MUSCLES OF THE TRUNK 71

CHAPTER SIX MUSCLES OF THE SHOULDER AND ARM 97

CHAPTER SEVEN MUSCLES OF THE FOREARM AND HAND 121

CHAPTER EIGHT MUSCLES OF THE HIP AND THIGH 155

CHAPTER NINE MUSCLES OF THE LEG AND FOOT 185

PREFACE

This book is a study guide and reference for the anatomy and actions of human skeletal muscles. It is designed for use by students of anatomy, physical education and health-related fields. It also serves as a compact reference for the practicing professional.

The first chapter presents labeled line drawings of the skeleton, which include all structures that are used in describing origins and insertions in the later chapters. A master numbering system is used so that each structure is labeled with the same number in all drawings.

The second chapter describes the various movements of the body.

In chapters 3 through 9 the origin, insertion, action and innervation of the skeletal muscles are described and each muscle is presented on a separate page with a line drawing.

The spinal cord levels of the nerve fibers that innervate each muscle are included in parentheses after the name of each nerve.

Labeled drawings of major muscle groups are presented throughout chapters 3 to 9. Notes and relationships among muscles have been included on many pages.

The drawings include the following important features:

1. Bones and cartilage containing muscle attachment sites are shaded.
2. Adjacent structures are shown.
3. Muscle fibers are drawn by direction.
4. Muscle fibers are shown on the undersurface of bone and cartilage as dashed lines.
5. Tendons and aponeuroses are shown.
6. Labeled muscle groups are included.

These features aid in visual orientation and understanding of the action of the muscles. We have noticed that many students find it useful to color the illustrations.

Notes have been included on many pages to show how muscles are used. Relationships among many of the muscles have also been indicated where appropriate. Many more of these have been included in the third edition.

Since our primary goal is to describe the muscles moving the skeleton, we have not described the muscles of the perineum, eye, tympanic cavity, tongue, larynx, pharynx, or palate.

We extend our appreciation to Mr. George Boykin, who was for many years the jolly proprietor of the gross anatomy laboratories at the Health Sciences Center of the State University of New York at Stony Brook, for his help and encouragement. We also thank Mr. Vincent Verdisco and Ms. Diane Chandler for their technical advice with the artwork and Ms. Katherine Juner for her secretarial services.

Robert J. Stone

Judith A. Stone

List of Reviewers

Steven G. Bassett
Seton Hill College

Wendy D. Bircher
San Juan College

Dianna L. Bourke
Penn State–Hazleton Campus

Linda Carpenter
Brooklyn College, Emeritus

Chris Chabot
Plymouth State College

Mitchell L. Cordova
Indiana State University

David T. Deutsch
School of Exercise, Leisure, and Sport
Kent State University

Christopher Dunbar
Brooklyn College, C.U.N.Y.

Rafael Escamilla
California Polytechnic State University

Nancy L. Goodyear
Bainbridge College

Jeannette K. Hafey
Springfield College

S. Taseer Hussain
Howard University College of Medicine

Roger L. Lane
Kent State University, Ashtabula Campus

Murray K. Marks
University of Tennessee

John Miller
University of New Hampshire

Elizabeth A. Murray
College of Mount St. Joseph

Joseph B. Myers
The University of North Carolina at Chapel Hill

David Pearson
Ball State University

Diane Redman
Sheridan College

Peter J. Wilkin
Purdue University North Central

CHAPTER ONE
THE SKELETON

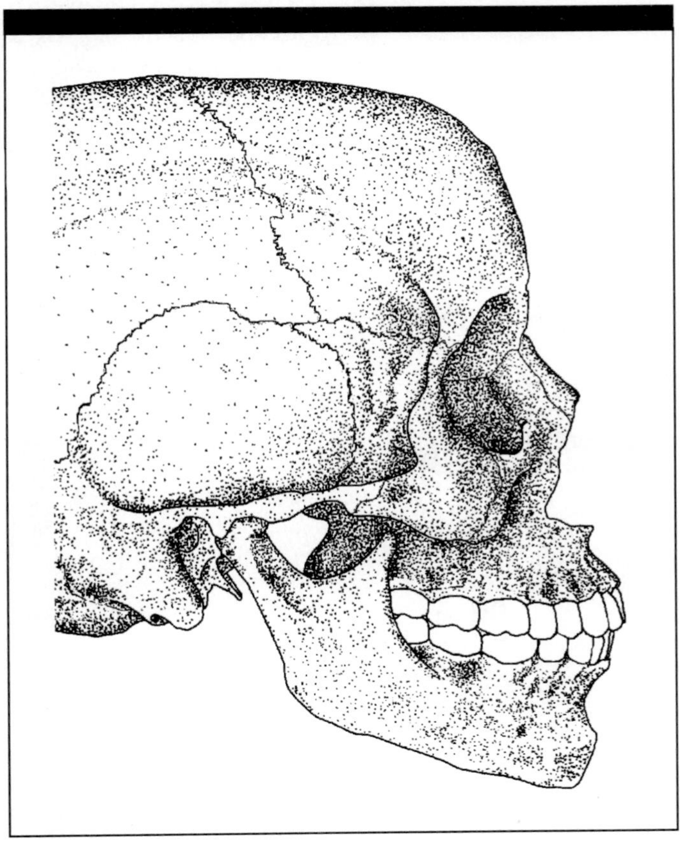

SKULL—LATERAL VIEW

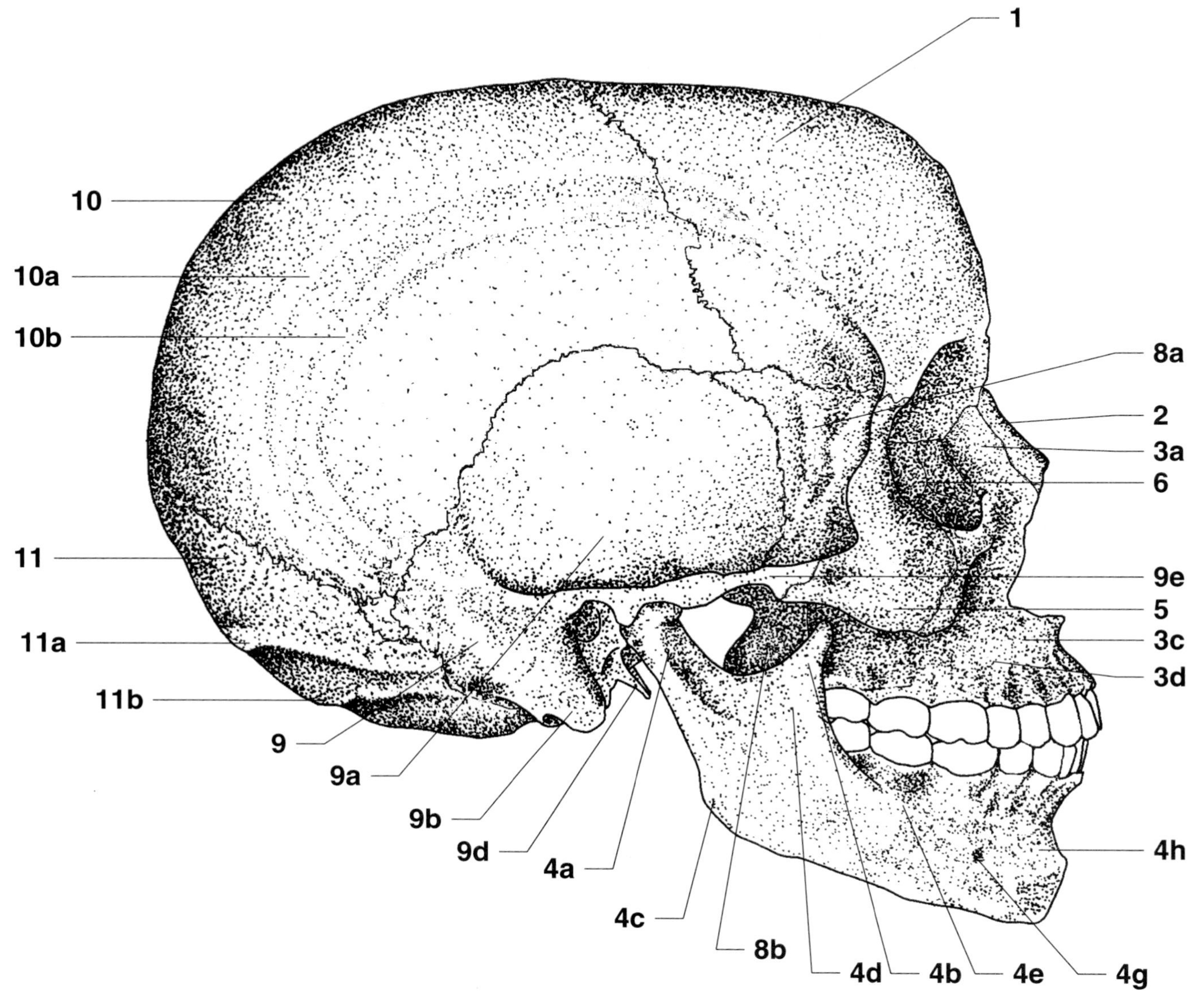

1. Frontal bone
2. Nasal bone
3a. Frontal process (maxilla)
3c. Incisive fossa of maxilla
3d. Canine fossa (maxilla)
4a. Neck of condyle (mandible)
4b. Coronoid process (mandible)
4c. Angle of the mandible
4d. Ramus (mandible)
4e. Oblique line (mandible)
4g. Mental foramen (mandible)
4h. Incisive fossa of mandible
5. Zygomatic bone
6. Lacrimal bone
8a. Greater wing of sphenoid bone
8b. Lateral pterygoid plate
9. Temporal bone
9a. Temporal fossa
9b. Mastoid process (temporal bone)
9d. Styloid process (temporal bone)
9e. Zygomatic process (temporal bone)
10. Parietal bone
10a. Superior temporal line
10b. Inferior temporal line
11. Occipital bone
11a. Superior nuchal line (occipital bone)
11b. Inferior nuchal line (occipital bone)

Note: The zygomatic arch is formed by the zygomatic process of the temporal bone meeting the zygomatic bone.

SKULL—LATERAL VIEW

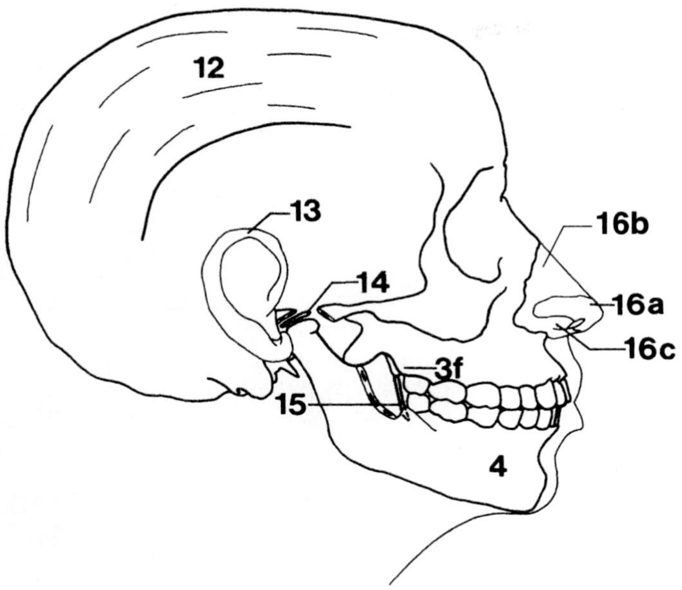

3f. Tuberosity of maxilla
4. Mandible
12. Galea aponeurotica
13. Helix of ear
14. Articular disk of temporomandibular joint
15. Pterygomandibular raphe
16a. Greater alar cartilage
16b. Nasal cartilage
16c. Ala

SKULL—POSTERIOR VIEW

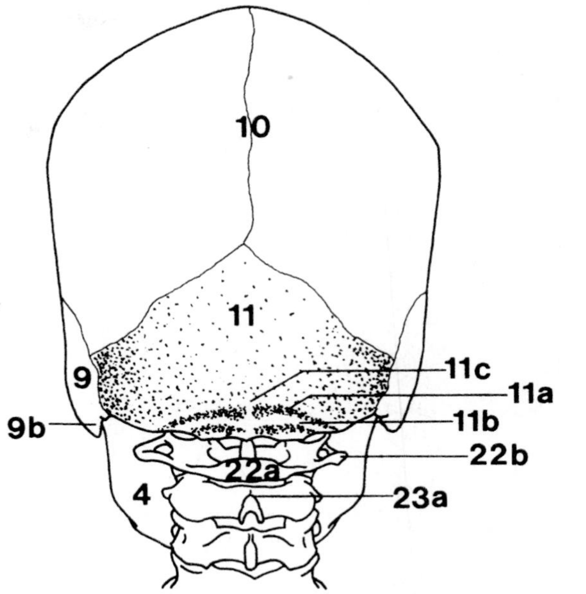

4. Mandible
9. Temporal bone
9b. Mastoid process (temporal bone)
10. Parietal bone
11. Occipital bone
11a. Superior nuchal line (occipital bone)
11b. Inferior nuchal line (occipital bone)
11c. External occipital protuberance
22a. Posterior arch of atlas
22b. Transverse process of atlas
23a. Spinous process of axis

SKULL—ANTERIOR VIEW

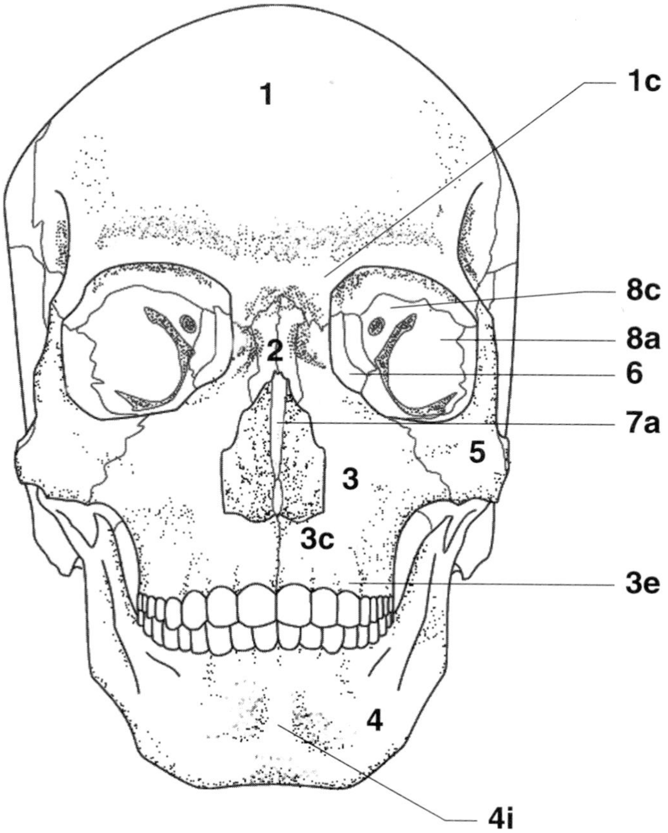

SKULL—INFERIOR (BASAL) VIEW

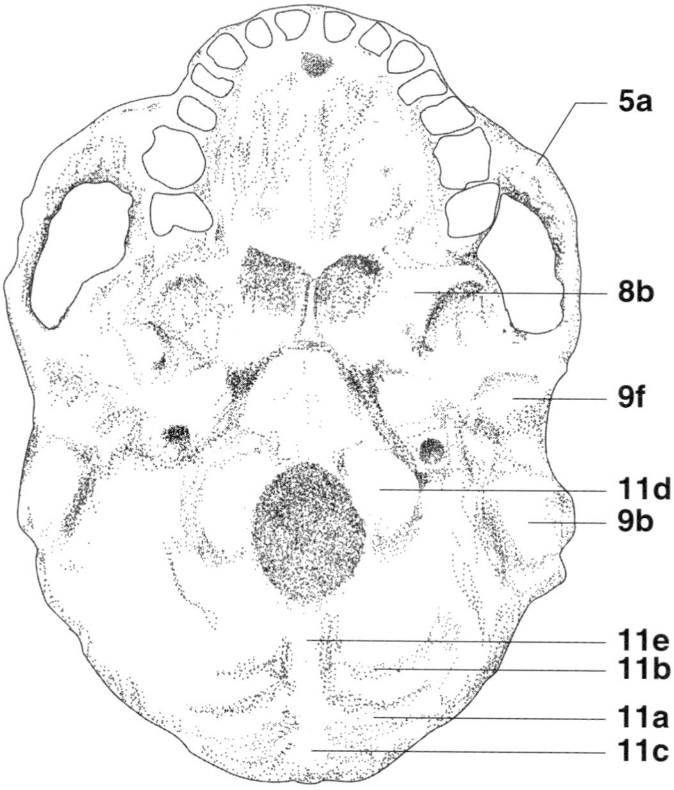

1. Frontal bone
1c. Superciliary arch (frontal bone)
2. Nasal bone
3. Maxilla
3c. Incisive fossa of maxilla
3e. Alveolar border of maxilla
4. Mandible
4i. Symphysis of mandible
5. Zygomatic bone
6. Lacrimal bone
7a. Nasal septum (ethmoid bone)
8a. Greater wing of sphenoid bone
8c. Lesser wing of sphenoid bone

5a. Zygomatic arch
8b. Lateral pterygoid plate
9b. Mastoid process (temporal bone)
9f. Mandibular process (temporal bone)
11a. Superior nuchal line (occipital bone)
11b. Inferior nuchal line (occipital bone)
11c. External occipital protuberance (occipital bone)
11d. Occipital condyle (occipital bone)
11e. External occipital crest (occipital bone)

SKULL TO HUMERUS—LATERAL VIEW

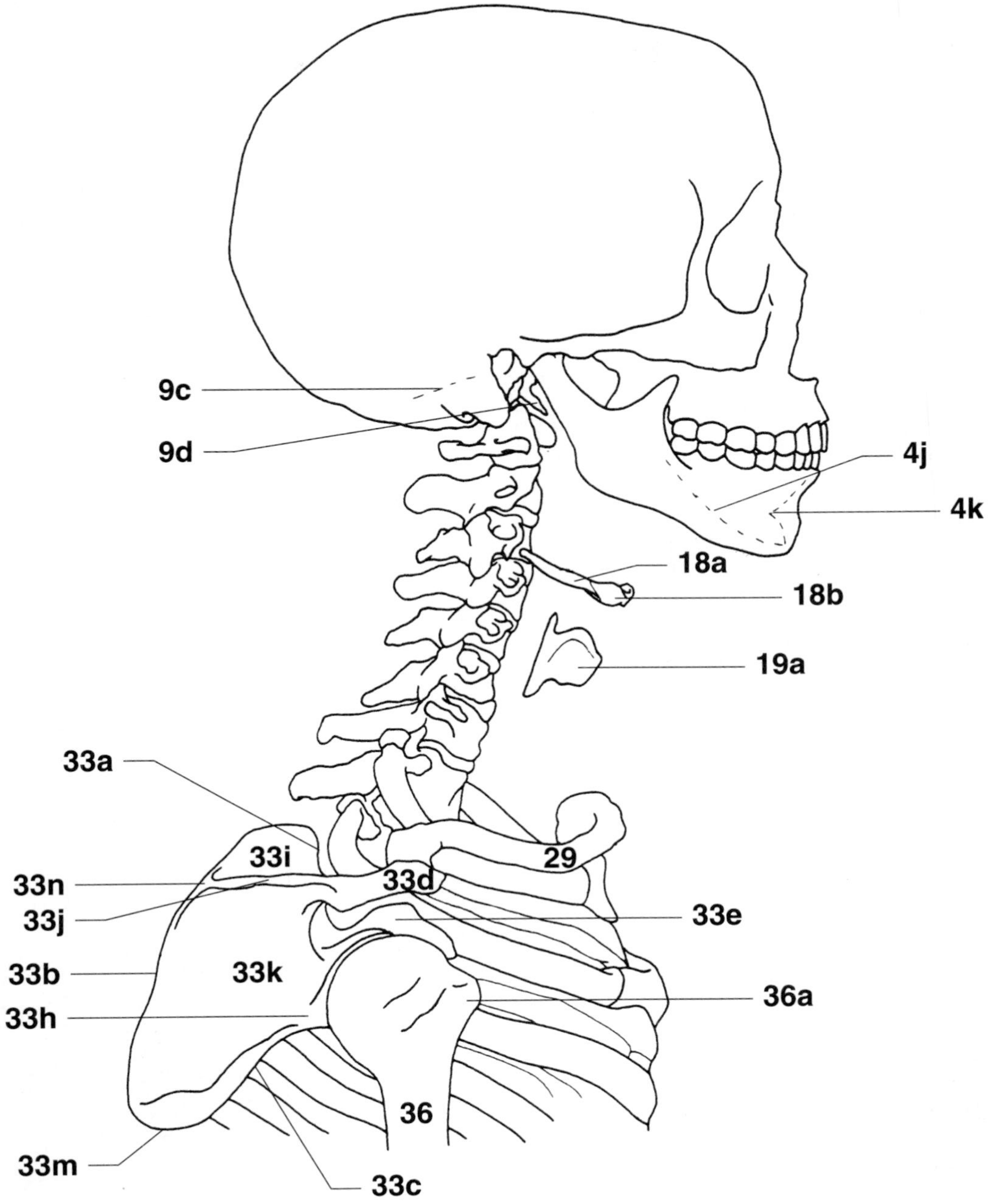

4j. Mylohyoid line (medial surface of mandible)
4k. Inferior mental spine (inner surface of mandible)
9c. Mastoid notch (medial surface of temporal bone)
9d. Styloid process (temporal bone)
18a. Greater cornu of hyoid
18b. Body of hyoid
19a. Lamina of thyroid cartilage
29. Clavicle
33a. Superior border of scapula
33b. Vertebral (medial) border of scapula
33c. Axillary (lateral) border of scapula
33d. Acromion (scapula)
33e. Coracoid process (scapula)
33h. Infraglenoid tubercle (scapula)
33i. Supraspinous fossa (scapula)
33j. Crest of spine (scapula)
33k. Infraspinous fossa (scapula)
33m. Inferior angle of scapula
33n. Root of spine (scapula)
36. Humerus
36a. Greater tuberosity of humerus

SKULL TO STERNUM—ANTERIOR VIEW

(Mandible and maxilla removed)

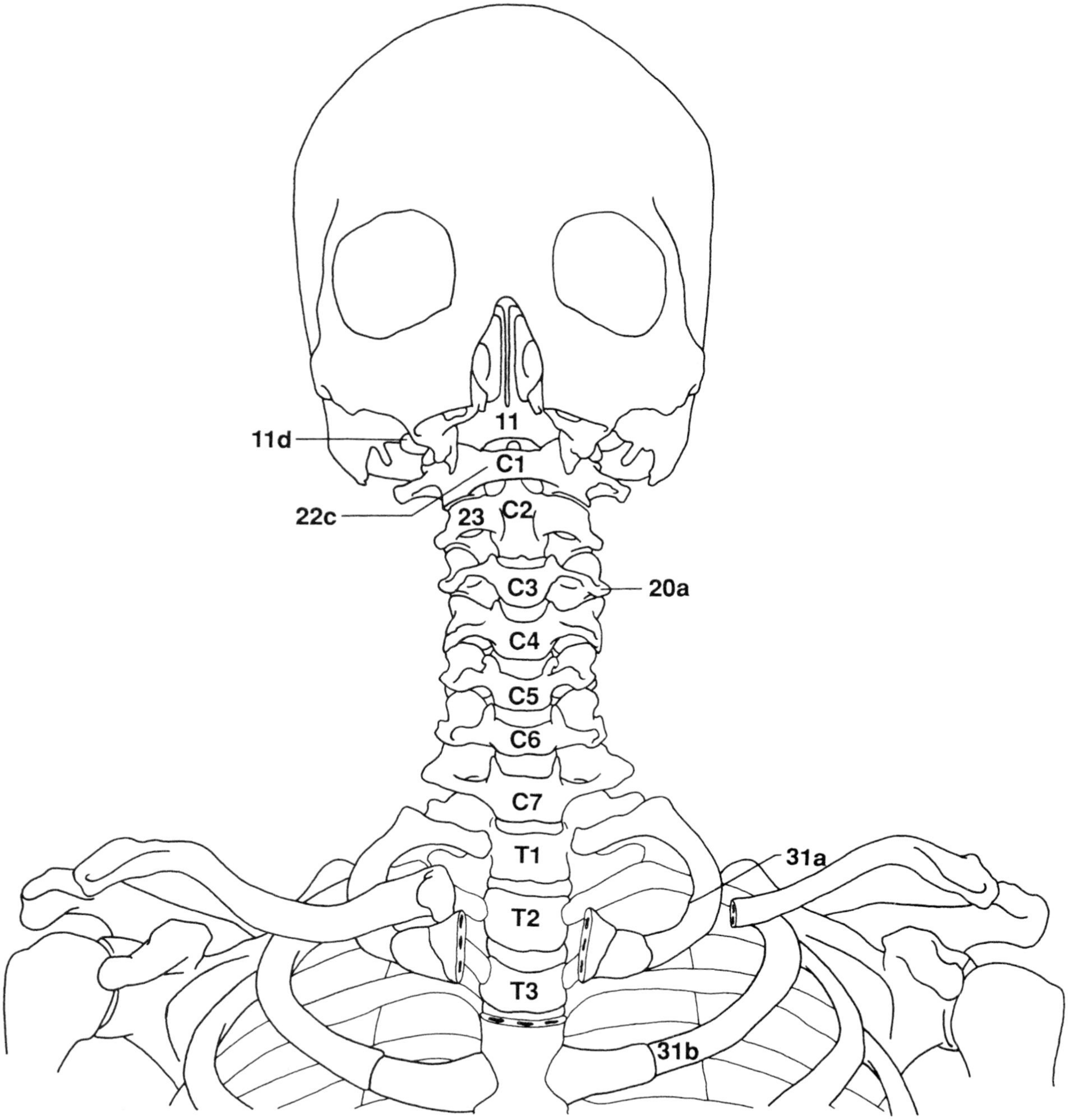

11. Occipital bone
11d. Jugular process of occipital bone
20a. Transverse process of vertebra
22c. Anterior arch of atlas
23. Axis
31a. Scalene tubercle of first rib
31b. Second rib

RIB CAGE, PECTORAL GIRDLE, UPPER ARM—ANTERIOR VIEW

(Ribs partially removed, right arm disarticulated)

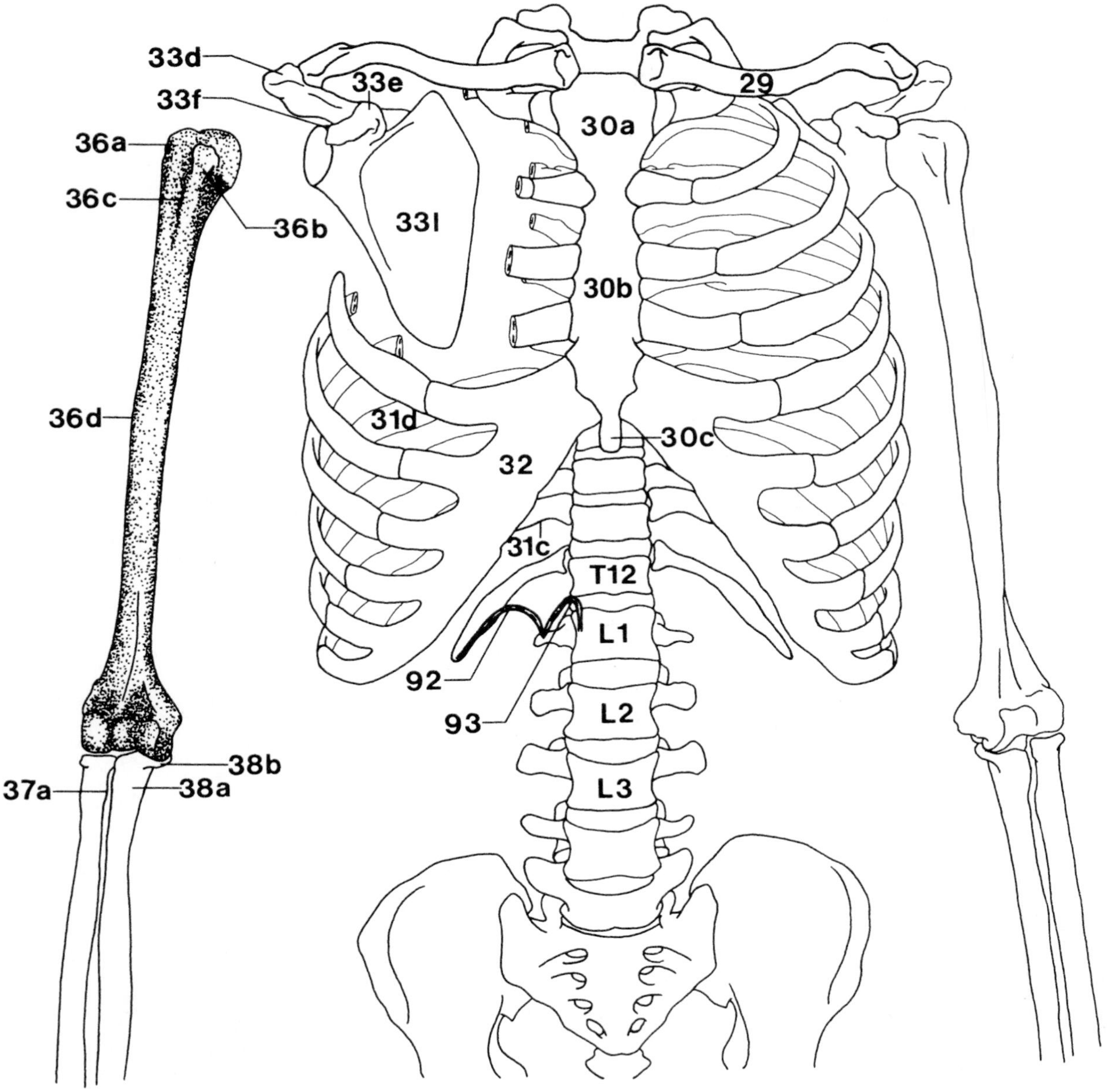

29. Clavicle
30a. Manubrium (sternum)
30b. Body (sternum)
30c. Xiphoid process (sternum)
31c. Tubercle of rib
31d. Angle of rib
32. Costal cartilage
33d. Acromion (scapula)
33e. Coracoid process (scapula)
33f. Supraglenoid tubercle (scapula)
33l. Subscapular fossa (scapula)
36a. Greater tuberosity (tubercle) of humerus
36b. Lesser tuberosity of humerus
36c. Intertubercular (bicipital) groove (humerus)
36d. Deltoid tuberosity (humerus)
37a. Radial tuberosity (radius)
38a. Ulnar tuberosity (ulna)
38b. Coronoid process (ulna)
92. Lateral lumbocostal arch (lateral arcuate ligament)*
93. Medial lumbocostal arch (medial arcuate ligament)

*See p. 89.

SKELETON—POSTERIOR VIEW

(Enlargement of lumbar vertebrae)

20a. Transverse process of vertebra
20b. Spinous process of vertebra
20c. Mamillary process of vertebra
20d. Accessory process of vertebra
24. Ligamentum nuchae
25. Supraspinous ligaments
26a. Lateral sacral crest
26b. Sacral foramina
26c. Medial sacral crest
29. Clavicle
31c. Tubercle of rib
31d. Angle of rib
33. Scapula
33d. Acromion (scapula)
36. Humerus
37. Radius
38. Ulna
38c. Olecranon process (ulna)
52a. Posterior superior iliac spine
52b. Iliac crest

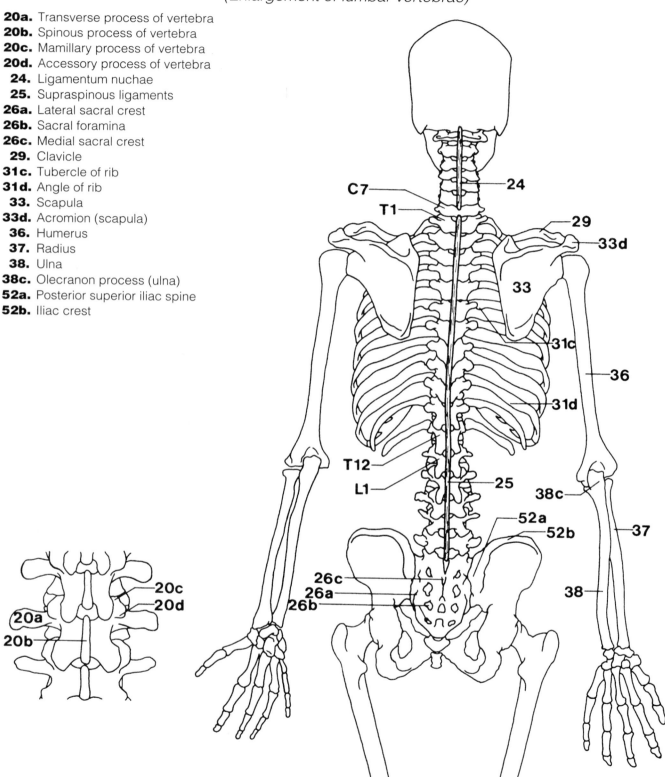

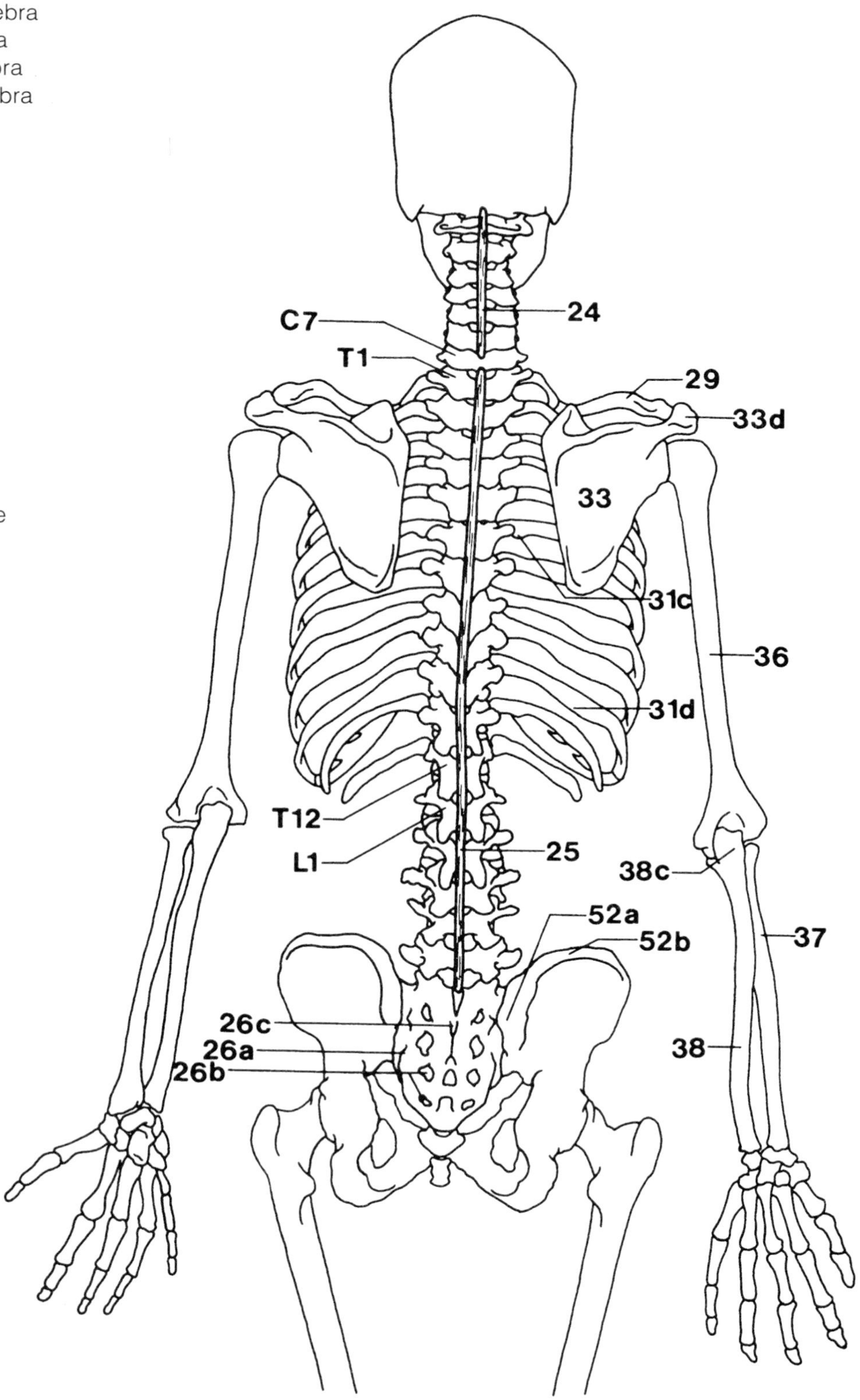

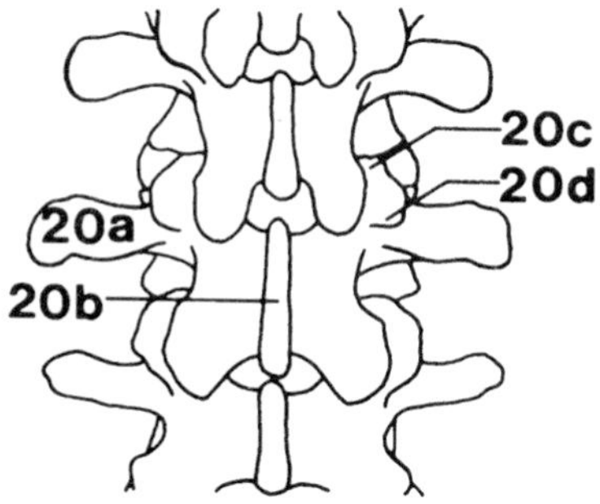

RIGHT ARM—POSTERIOR VIEW

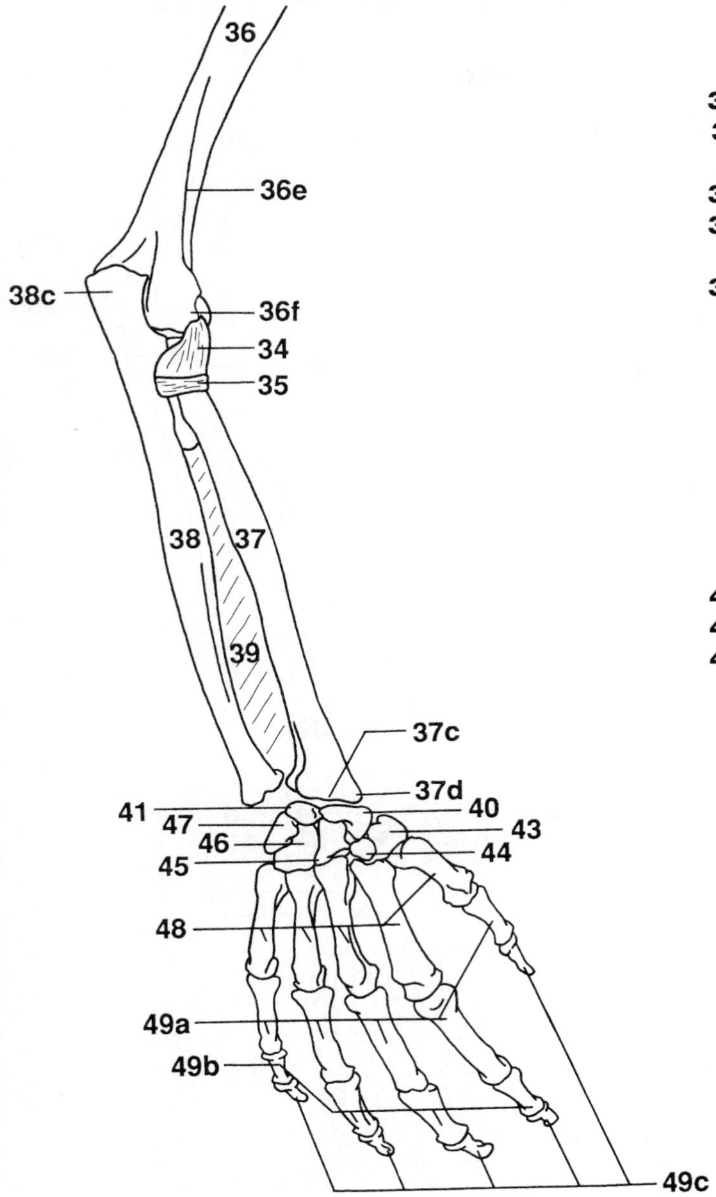

34. Radial collateral ligament
35. Annular ligament
36. Humerus
36e. Lateral supracondylar ridge (humerus)
36f. Lateral epicondyle (humerus)
37. Radius
37c. Dorsal tubercle (radius)
37d. Styloid process (radius)
38. Ulna
38c. Olecranon process (ulna)
39. Interosseous membrane
40. Scaphoid (navicular)
41. Lunate
43. Trapezium
44. Trapezoid
45. Capitate
46. Hamate
47. Triquetrum
48. Metacarpals
49a. Proximal phalanges
49b. Middle phalanges
49c. Distal phalanges

RIGHT HAND—ANTERIOR VIEW

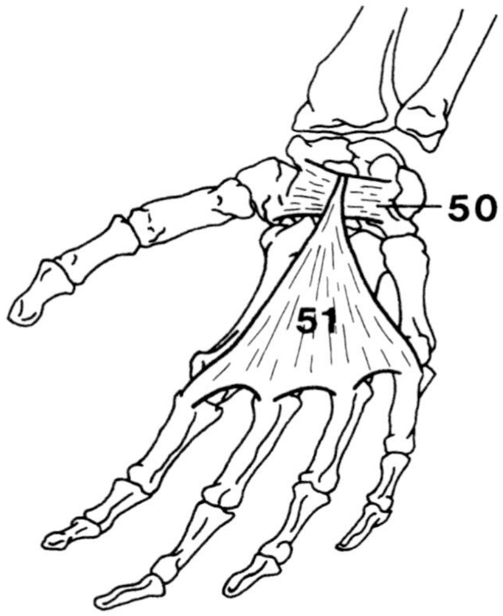

RIGHT ARM—ANTERIOR VIEW

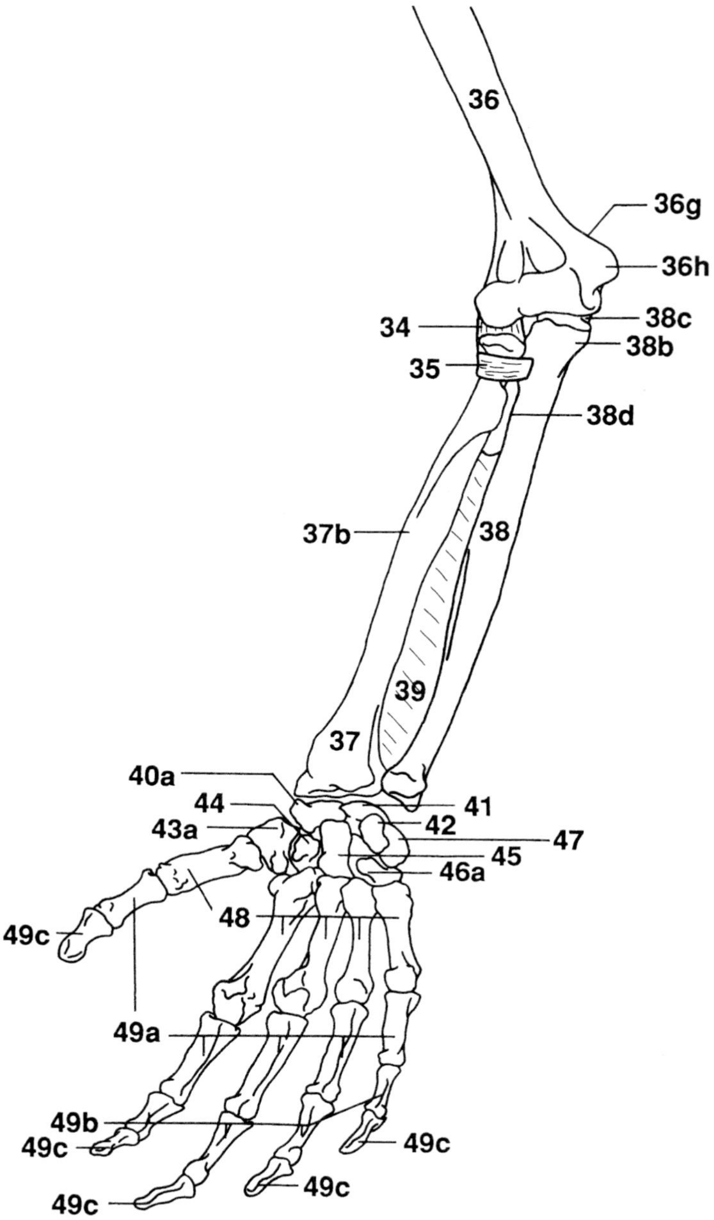

34. Radial collateral ligament
35. Annular ligament
36. Humerus
36g. Medial supracondylar ridge (humerus)
36h. Medial epicondyle (humerus)
37. Radius
37b. Pronator tuberosity (radius)
38. Ulna
38b. Coronoid process (ulna)
38c. Olecranon process (ulna)
38d. Supinator crest (ulna)
39. Interosseous membrane
40a. Tubercle of scaphoid (navicular)
41. Lunate
42. Pisiform
43a. Tubercle of trapezium
44. Trapezoid
45. Capitate
46a. Hook of hamate
47. Triquetrum
48. Metacarpals
49a. Proximal (first) phalanges
49b. Middle (second) phalanges
49c. Distal (third) phalanges
50. Flexor retinaculum
51. Palmar aponeurosis

LUMBAR AND PELVIC REGION—ANTERIOR VIEW

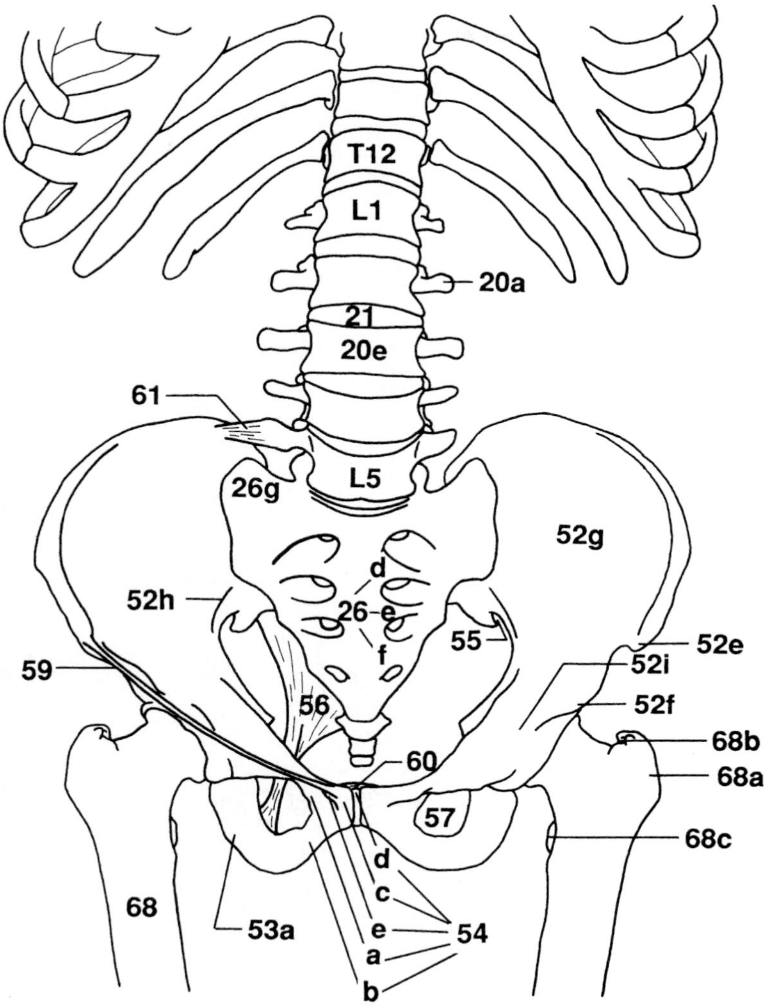

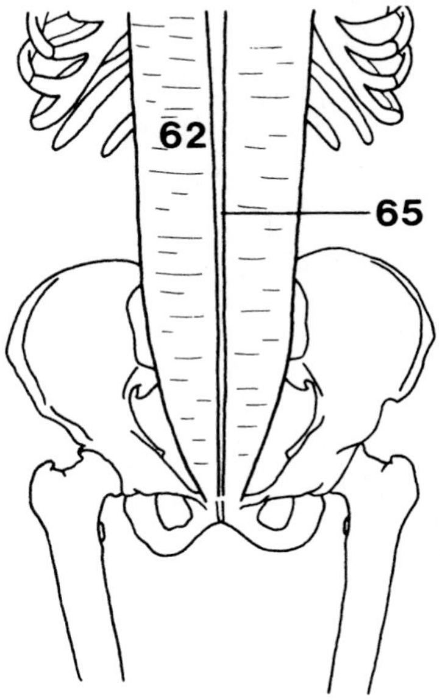

20a. Transverse process of vertebra
20e. Body of vertebra
21. Intervertebral disk
26d. Second sacral vertebra
26e. Third sacral vertebra
26f. Fourth sacral vertebra
26g. Ala of sacrum
52e. Anterior superior iliac spine
52f. Anterior inferior iliac spine
52g. Iliac fossa
52h. Arcuate line (ilium)
52i. Iliopectineal eminence (ilium)
53a. Ramus of ischium
54a. Superior ramus of pubis
54b. Inferior ramus of pubis
54c. Pubic crest
54d. Pubic symphysis
54e. Pubic tubercle
55. Greater sciatic notch
56. Sacrotuberous ligament
57. Obturator foramen
59. Inguinal ligament
60. Superior pubic ligament
61. Iliolumbar ligament
62. Rectus sheath
65. Linea alba
68. Femur
68a. Greater trochanter (femur)
68b. Trochanteric fossa (femur)
68c. Lesser trochanter (femur)

PELVIC GIRDLE TO KNEE—LATERAL VIEW

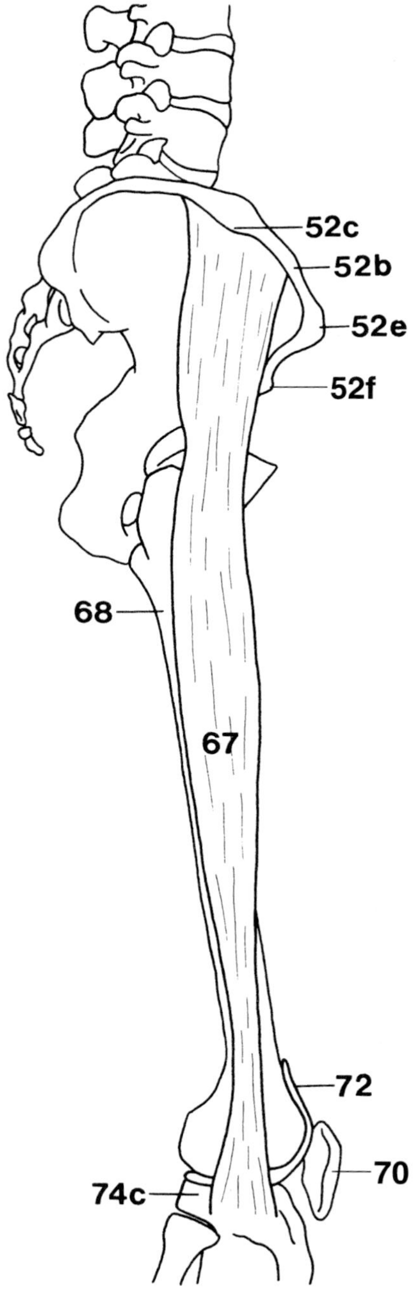

52b. Iliac crest
52c. Iliac tubercle
52e. Anterior superior iliac spine
52f. Anterior inferior iliac spine
67. Iliotibial tract
68. Femur
70. Patella
72. Synovial membrane of knee joint
74c. Lateral condyle of tibia

THORACIC TO PELVIC REGION—LATERAL VIEW

(Arm and leg removed)

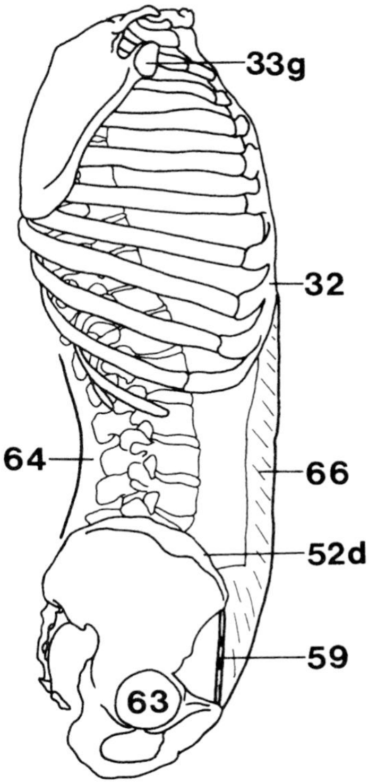

32. Costal cartilage
33g. Glenoid cavity (scapula)
52d. Anterior iliac crest
59. Inguinal ligament
63. Acetabulum
64. Thoracolumbar fascia
66. Abdominal aponeurosis

PELVIC GIRDLE—POSTERIOR VIEW

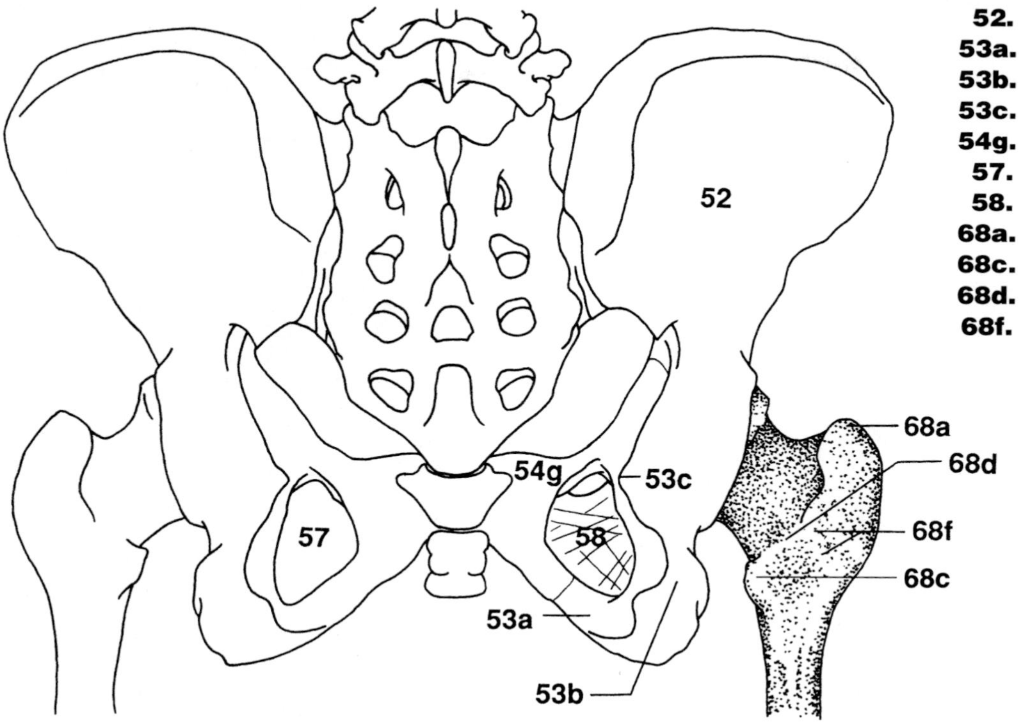

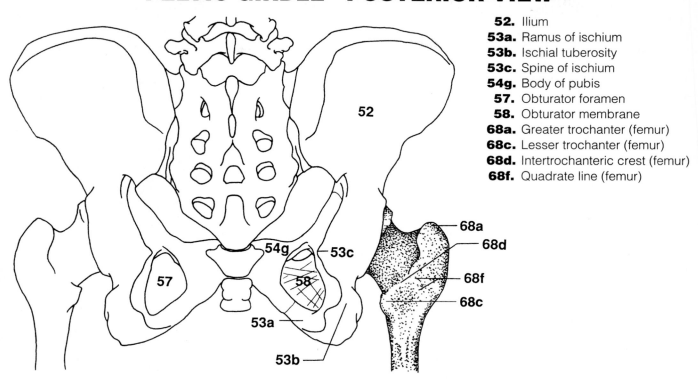

52. Ilium
53a. Ramus of ischium
53b. Ischial tuberosity
53c. Spine of ischium
54g. Body of pubis
57. Obturator foramen
58. Obturator membrane
68a. Greater trochanter (femur)
68c. Lesser trochanter (femur)
68d. Intertrochanteric crest (femur)
68f. Quadrate line (femur)

PELVIC GIRDLE AND UPPER LEG—THREE-QUARTER POSTERIOR VIEW

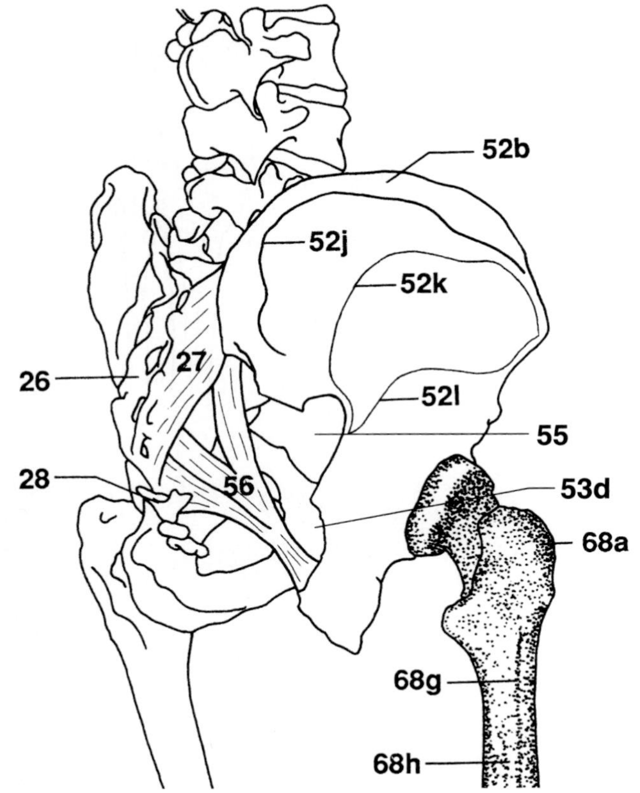

26. Sacrum
27. Aponeurosis of erector spinae
28. Coccyx
52b. Iliac crest
52j. Posterior gluteal line (ilium)
52k. Middle (anterior) gluteal line (ilium)
52l. Inferior gluteal line (ilium)
53d. Lesser sciatic notch
55. Greater sciatic notch
56. Sacrotuberous ligament
68a. Greater trochanter (femur)
68g. Gluteal tuberosity (femur)
68h. Linea aspera (femur)

PELVIC GIRDLE TO LEG—ANTERIOR VIEW

52. Ilium
52e. Anterior superior iliac spine
52f. Anterior inferior iliac spine
54b. Inferior ramus of pubis
54f. Pectineal line (pubis)
54g. Body of pubis
68. Femur
68a. Greater trochanter (femur)
68c. Lesser trochanter (femur)
68e. Intertrochanteric line (femur)
68i. Lateral supracondylar line (femur)
68j. Medial supracondylar line (femur)
69. Quadriceps tendon
70. Patella
71. Patellar ligament
74. Tibia
74a. Tuberosity of tibia

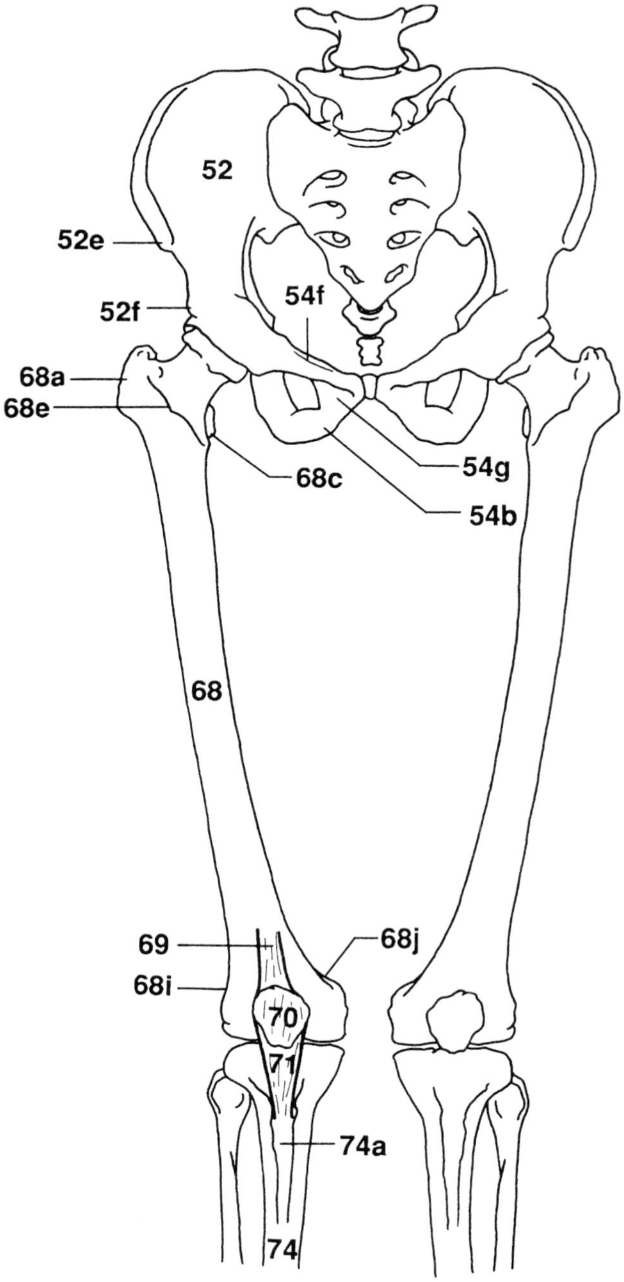

PELVIC GIRDLE TO LEG—POSTERIOR VIEW

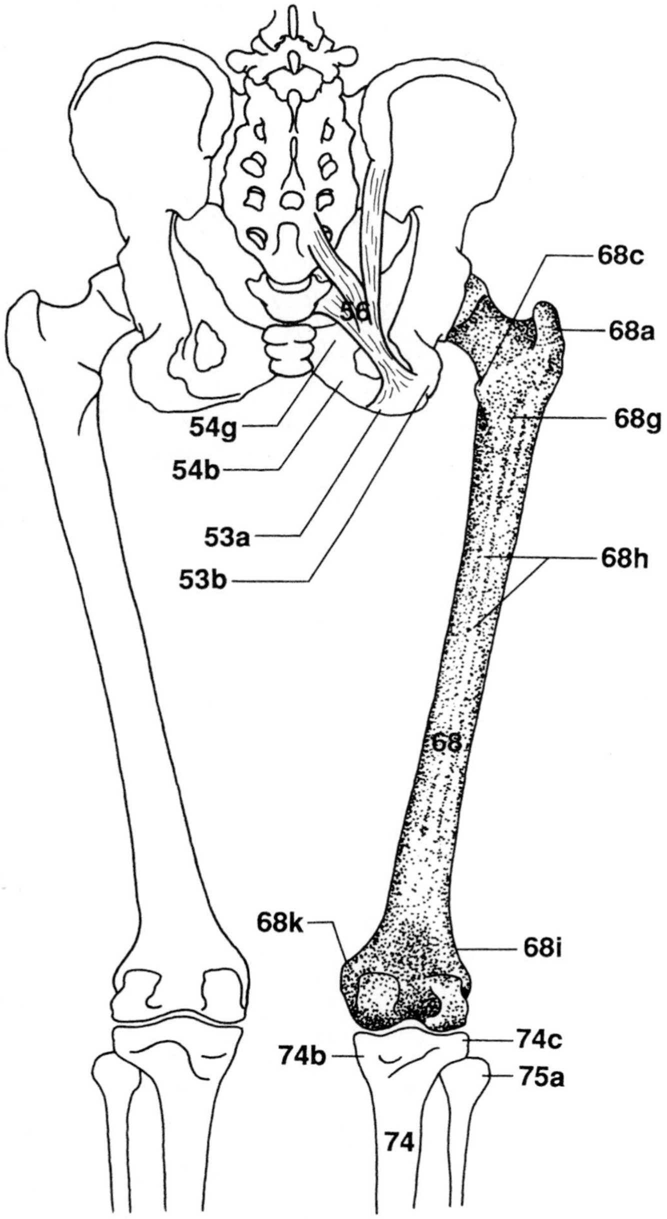

53a. Ramus of ischium
53b. Ischial tuberosity
54b. Inferior ramus of pubis
54g. Body of pubis
56. Sacrotuberous ligament
68. Femur
68a. Greater trochanter (femur)
68c. Lesser trochanter (femur)
68g. Gluteal tuberosity (femur)
68h. Linea aspera (femur)
68i. Lateral supracondylar line (femur)
68k. Adductor tubercle (femur)
74. Tibia
74b. Medial condyle of tibia
74c. Lateral condyle of tibia
75a. Head of fibula

RIGHT FOOT— ANTEROLATERAL VIEW

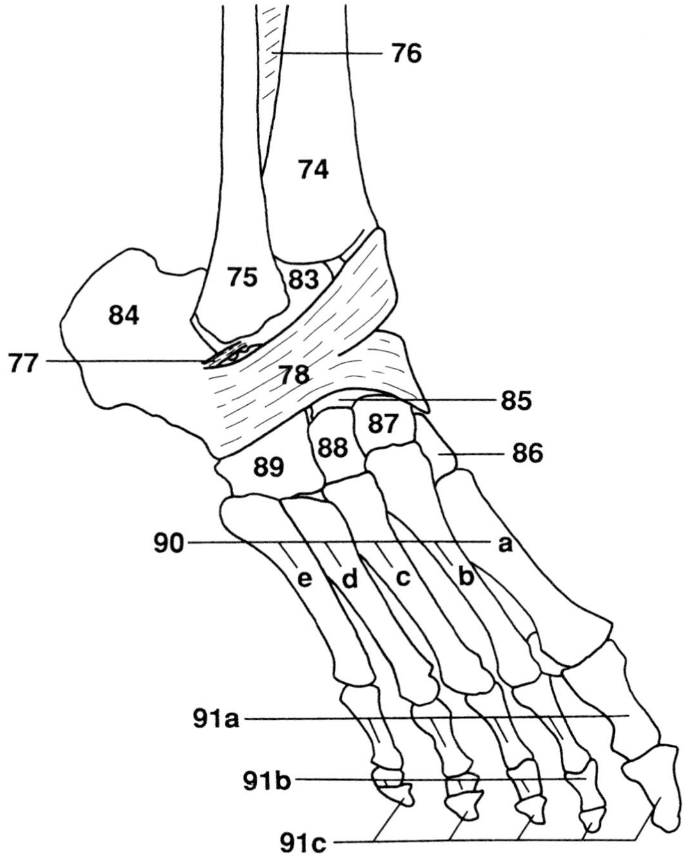

RIGHT LEG— ANTEROLATERAL VIEW

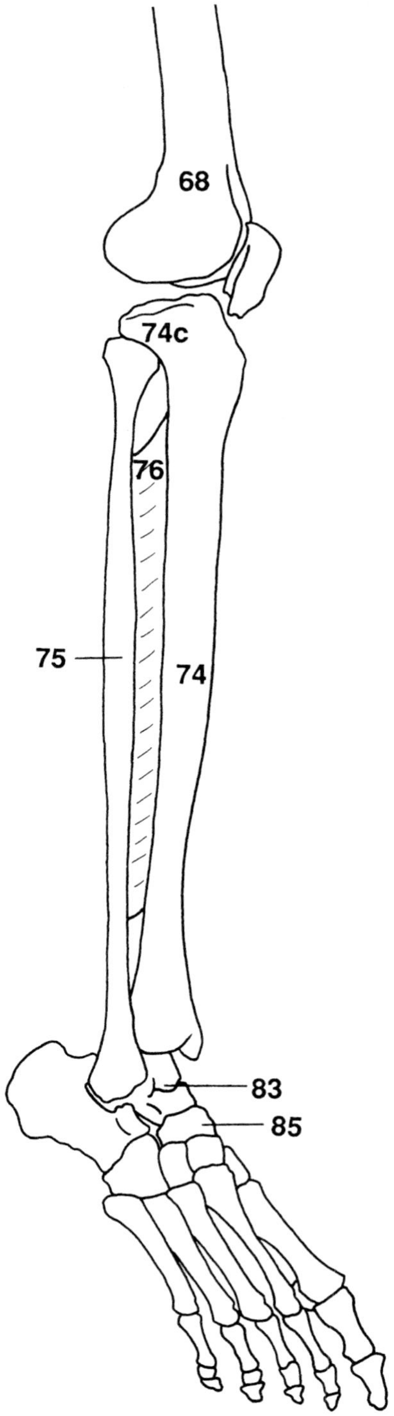

68. Femur
74. Tibia
74c. Lateral condyle of tibia
75. Fibula
76. Interosseous membrane
77. Lateral talocalcaneal ligament
78. Inferior extensor retinaculum
83. Talus
84. Calcaneus
85. Navicular
86. Medial cuneiform
87. Intermediate cuneiform
88. Lateral cuneiform
89. Cuboid
90. Metatarsal bones
90a. First metatarsal
90b. Second metatarsal
90c. Third metatarsal
90d. Fourth metatarsal
90e. Fifth metatarsal
91a. Proximal phalanges
91b. Middle phalanges
91c. Distal phalanges

RIGHT LEG—POSTERIOR VIEW

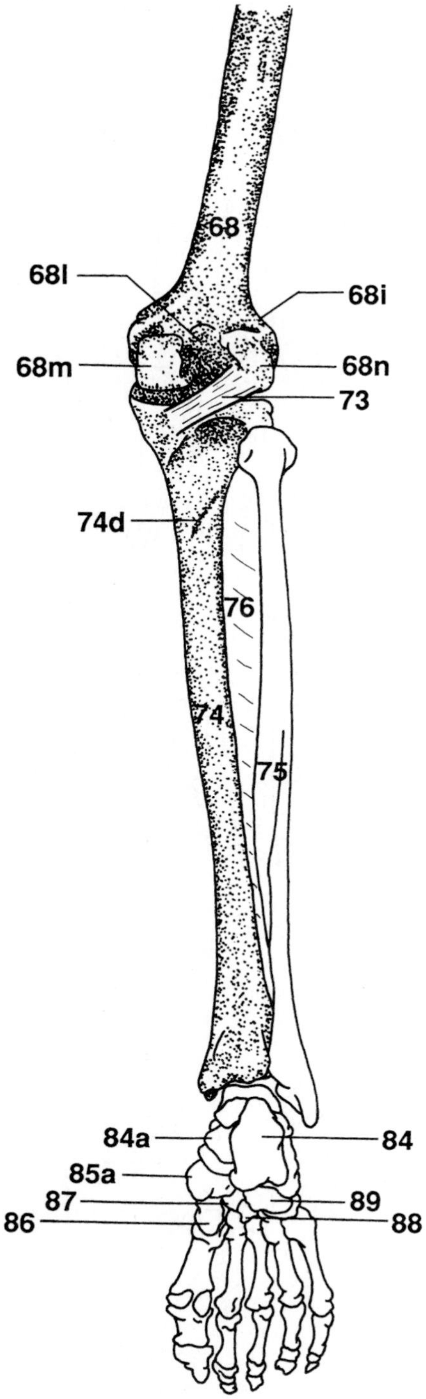

68. Femur
68i. Lateral supracondylar line (femur)
68l. Popliteal surface (femur)
68m. Medial condyle (femur)
68n. Lateral condyle (femur)
73. Oblique popliteal ligament
74. Tibia
74d. Soleal line (tibia)
75. Fibula
76. Interosseous membrane
84. Calcaneus
84a. Sustentaculum tali of calcaneus
85a. Tuberosity of navicular
86. Medial cuneiform
87. Intermediate cuneiform
88. Lateral cuneiform
89. Cuboid

RIGHT FOOT—PLANTAR VIEW

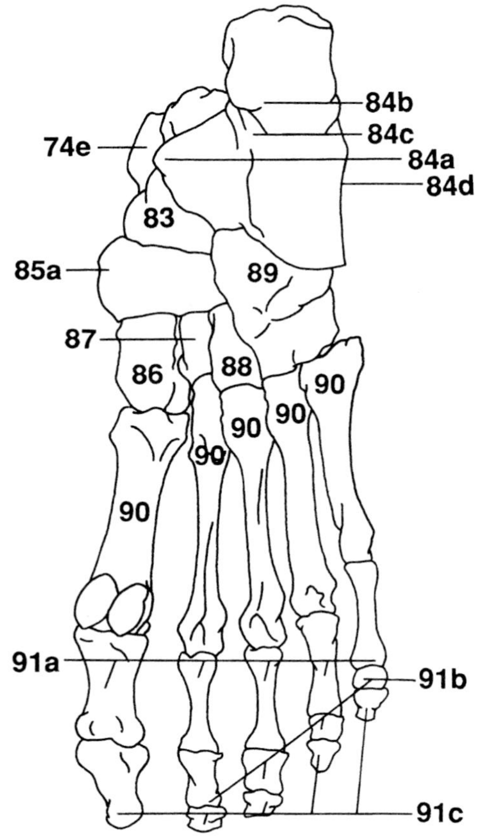

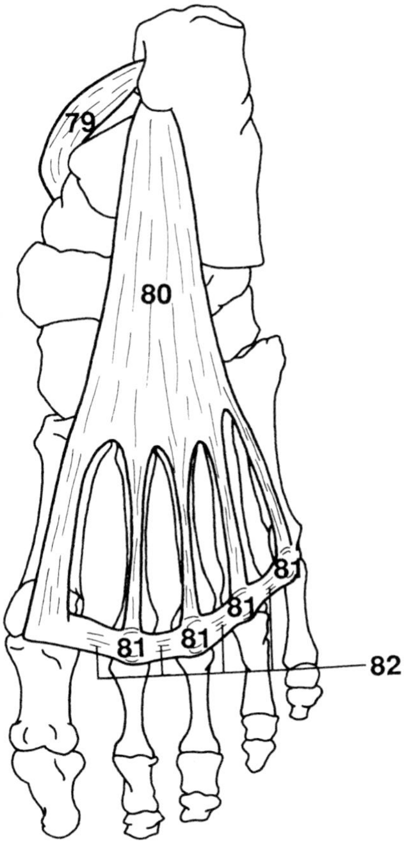

74e. Medial malleolus of tibia
79. Flexor retinaculum
80. Plantar aponeurosis
81. Plantar metatarsophalangeal ligaments
82. Transverse metatarsal ligaments
83. Talus
84a. Sustentaculum tali of calcaneus
84b. Tuberosity of calcaneus
84c. Medial border of calcaneus
84d. Lateral border of calcaneus
85a. Tuberosity of navicular
86. Medial cuneiform
87. Intermediate cuneiform
88. Lateral cuneiform
89. Cuboid
90. Metatarsal bones
91a. Proximal phalanges
91b. Middle phalanges
91c. Distal phalanges

CHAPTER TWO
MOVEMENTS OF THE BODY

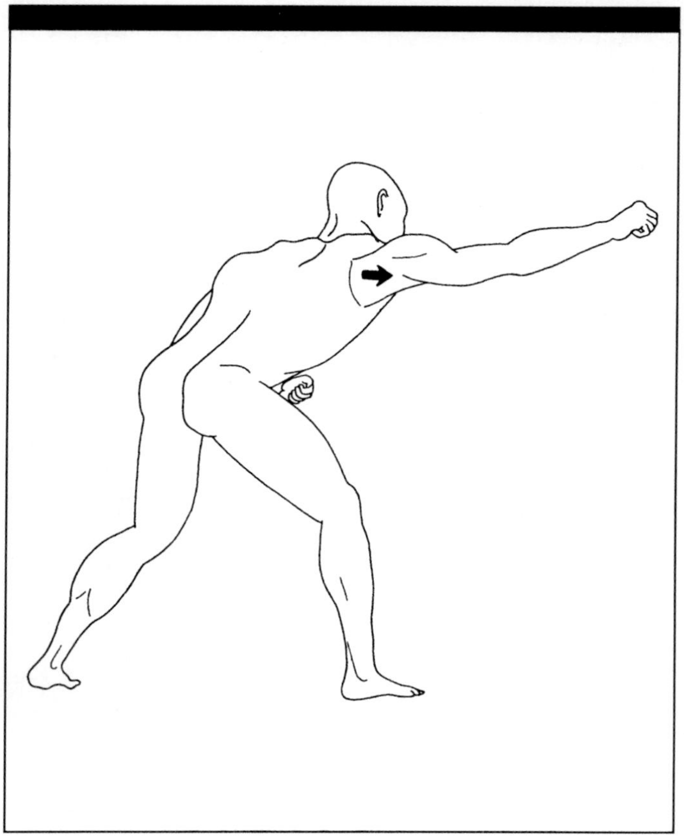

Anatomical position—A subject in the anatomical position is standing erect with the head, eyes, and toes facing forward and the arms hanging straight at the sides with the palms of the hands facing forward.

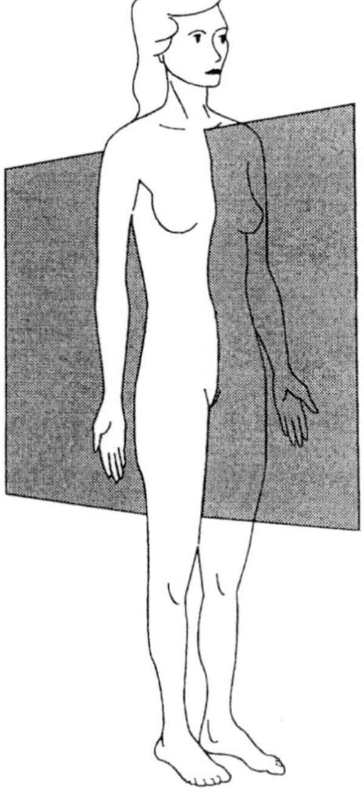

Figure 2.1
Median or midsagittal plane—Passes vertically through the body from anterior (front) to posterior (back). It divides the body into right and left sides. Other sagittal planes are parallel to this plane.

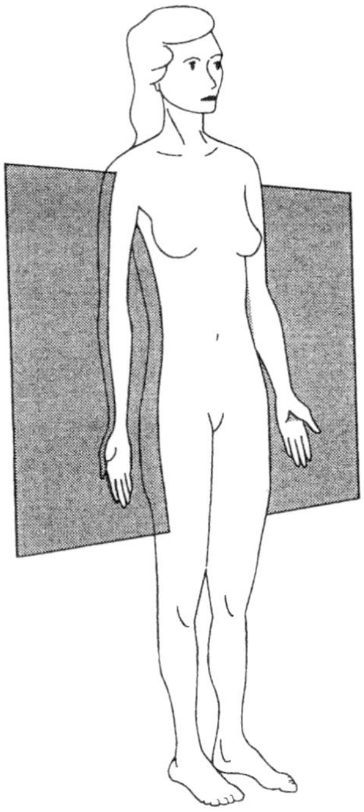

Figure 2.2
Coronal (frontal) planes—Pass vertically through the body from side to side. They divide the body from front to back.

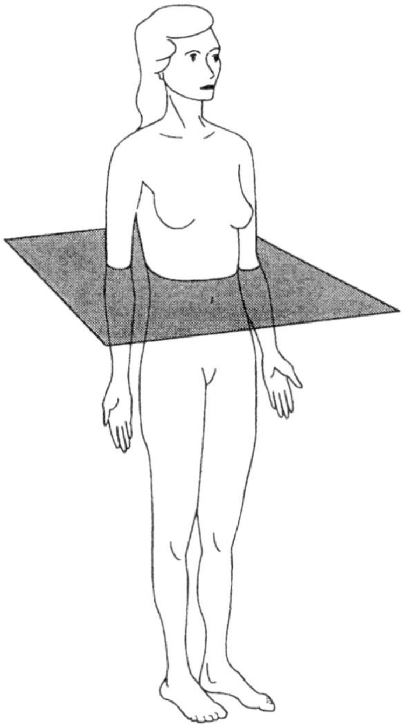

Figure 2.3
Transverse planes (cross sections)—Pass horizontally through the body parallel to the ground.

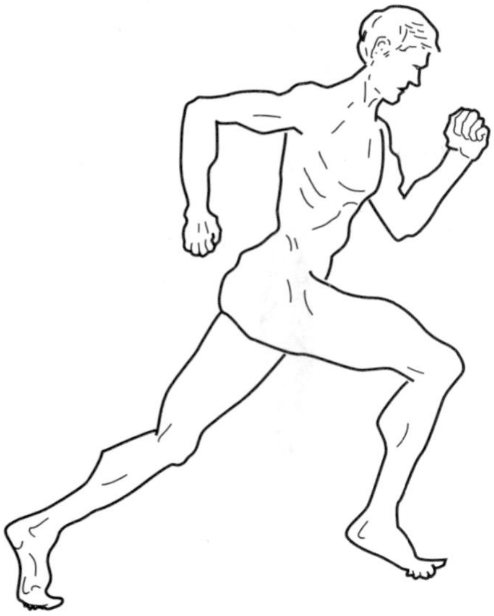

Figure 2.4
Flexion—The left arm, forearm, and right thigh are drawn forward in sagittal planes. The right knee is also flexed.
Extension—The left thigh and knee are extended.
Hyperextension—The right arm is hyperextended at the shoulder.

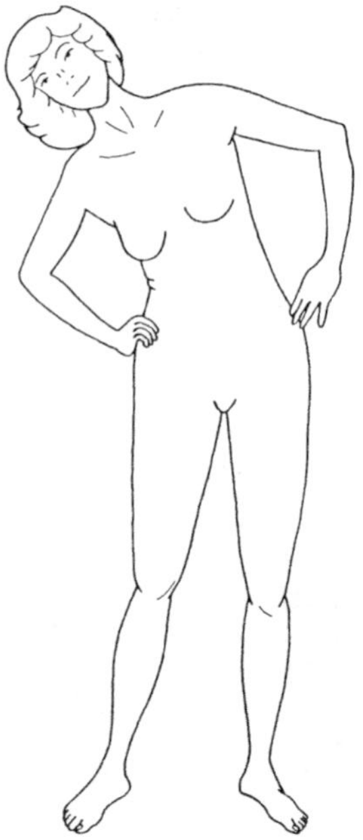

Figure 2.5
Lateral flexion—The torso (or head) bends laterally in the coronal plane.

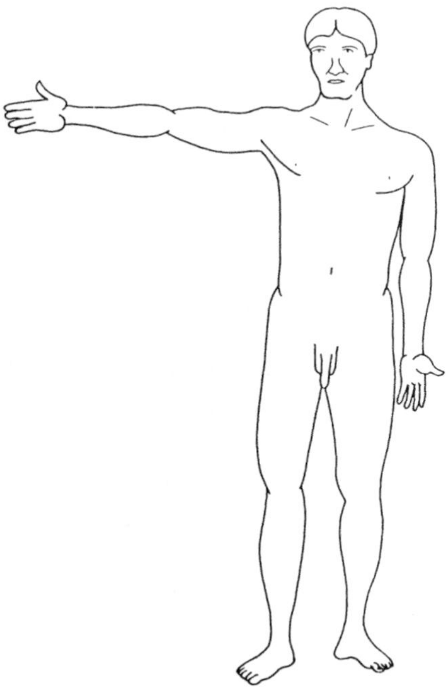

Figure 2.6
Abduction—The right arm is drawn laterally in the coronal plane.
Adduction—The left arm is returned from abduction to the anatomical position.

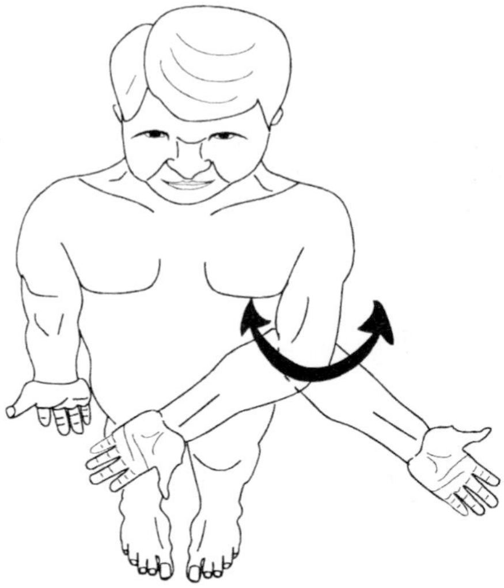

Figure 2.7
Medial rotation—The anterior of the arm (or thigh) is moved toward the median plane.
Lateral rotation—The anterior of the arm (or thigh) is moved away from the median plane.

MOVEMENTS OF THE SCAPULA

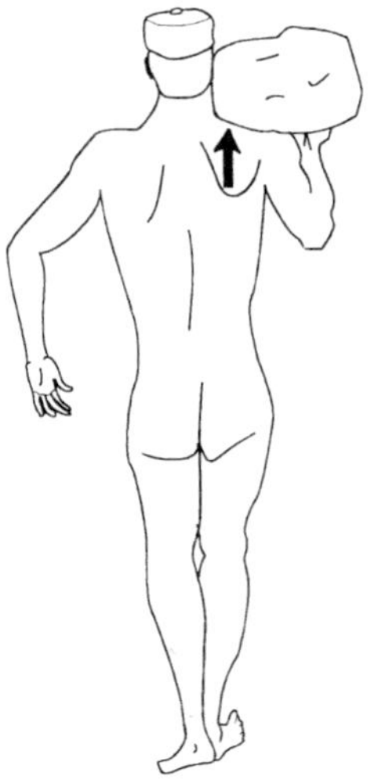

Figure 2.8
Elevation—The right scapula of this figure is drawn superiorly.

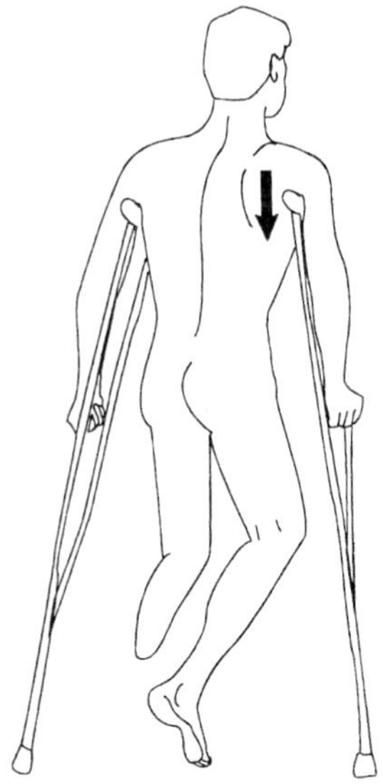

Figure 2.9
Depression—The right scapula of this figure is pushing the arm inferiorly.

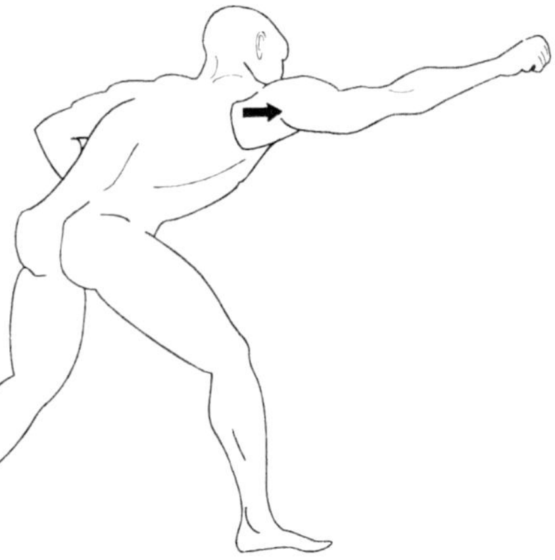

Figure 2.10
Protraction—The scapula pushes the arm forward in a sagittal plane.

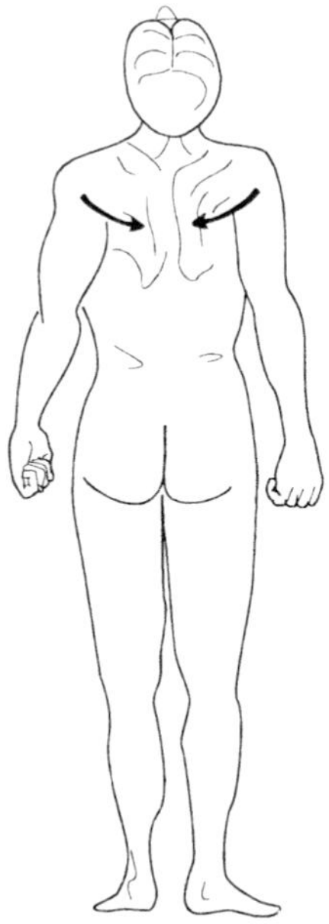

Figure 2.11
Retraction—The scapula is pulled back from protraction in a sagittal plane. Since the scapula slides around the ribs toward the median plane, it becomes adduction.

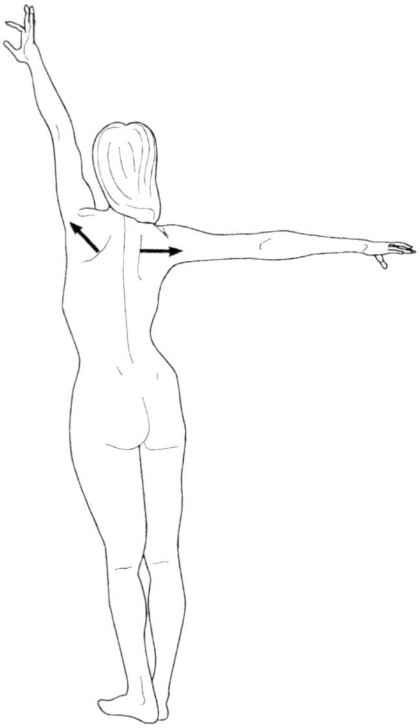

Figure 2.12
Rotation—For abduction of the arm to continue above the height of the shoulder, the scapula must rotate on its axis so that the glenoid fossa turns upward.

MOVEMENTS OF THE HAND AND FOREARM

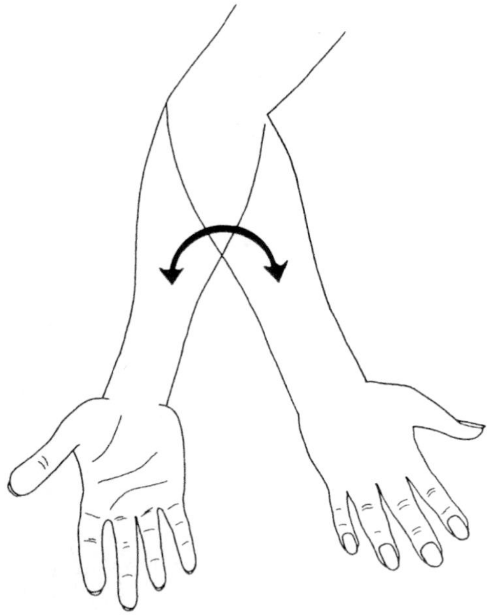

Figure 2.13
Pronation—The forearm is rotated away from the anatomical position so that the palm turns medially then posteriorly. If the forearm is flexed at the elbow, then the palm turns inferiorly.
Supination—The forearm is rotated so that the palm turns anteriorly (or superiorly if the forearm is flexed).

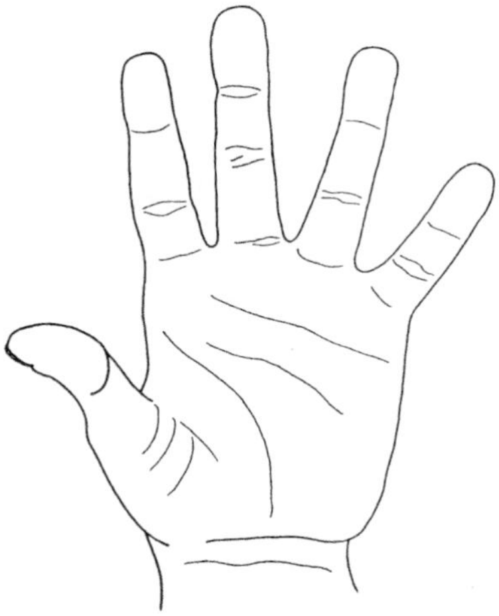

Figure 2.14
Abduction—The fingers are moved away from the midline of the hand.

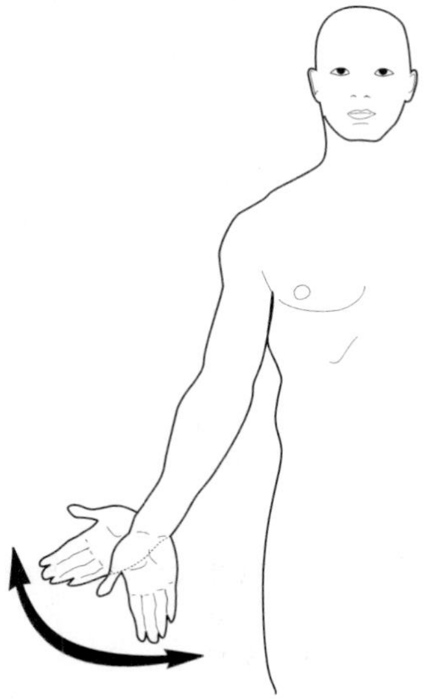

Figure 2.15
Radial flexion (abduction)—The hand, at the wrist, is drawn away from the body in a coronal plane.
Ulnar flexion (adduction)—The hand, at the wrist, is drawn toward the body in a coronal plane.

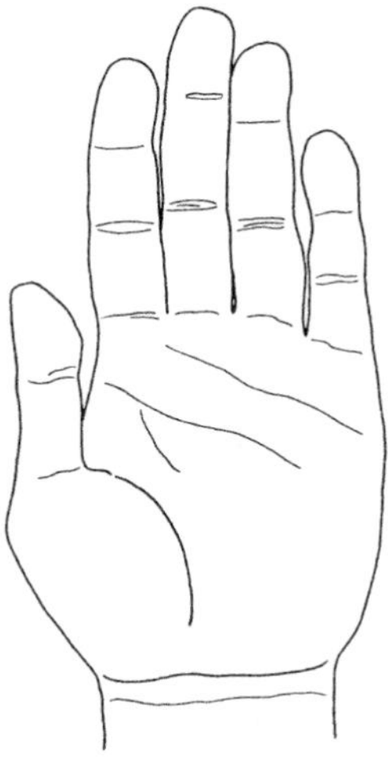

Figure 2.16
Adduction—The fingers are moved toward the midline of the hand.

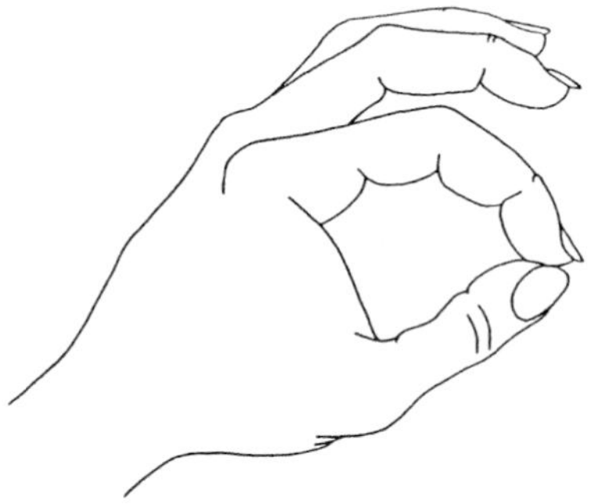

Figure 2.17
Opposition—The thumb is rotated so its anterior pad can touch the anterior pads of the four fingers.

MOVEMENTS OF THE FOOT

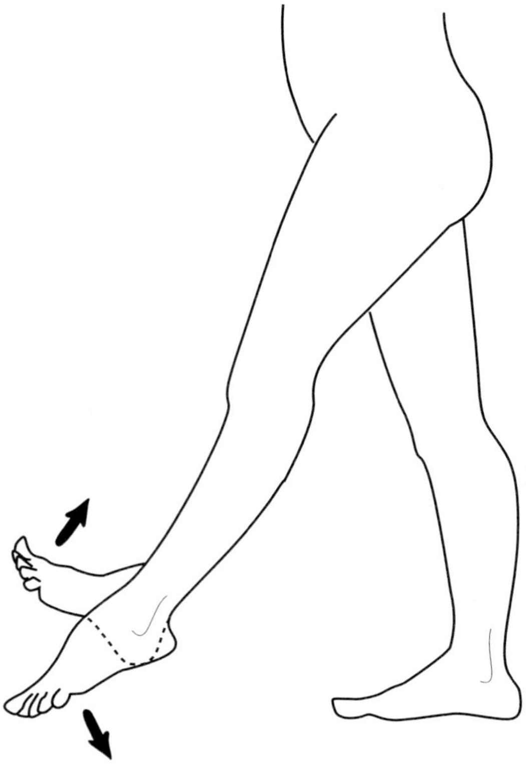

Figure 2.18
Dorsiflexion—The ankle flexes, moving the foot superiorly. Plantar Flexion—The ankle extends, moving the foot inferiorly.

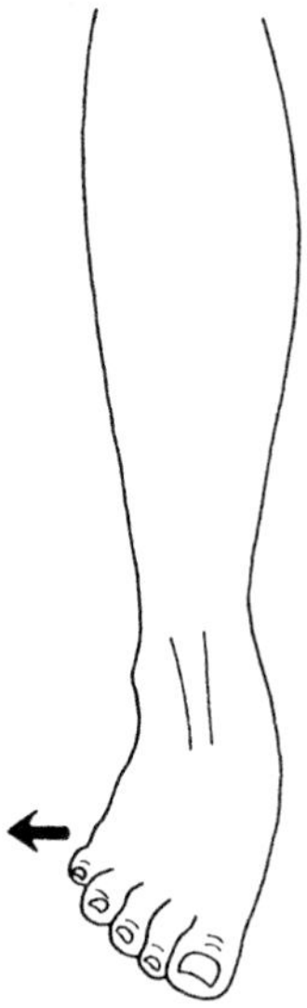

Figure 2.19
Eversion—The front of the foot moves laterally away from the midline (abduction), and the sole turns outward.

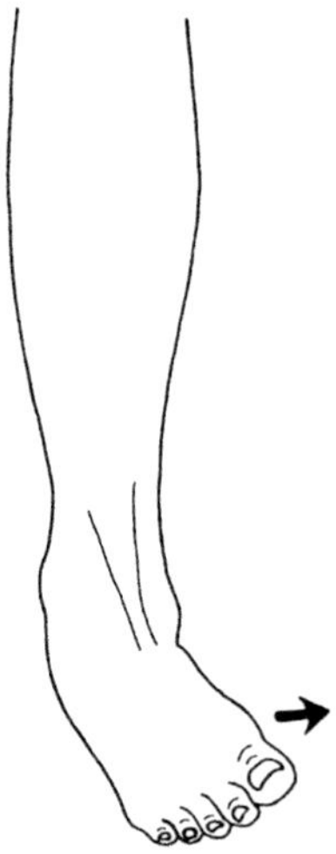

Figure 2.20
Inversion—The front of the foot moves medially toward the midline (adduction), and the sole turns inward.

CHAPTER THREE

MUSCLES OF THE FACE AND HEAD

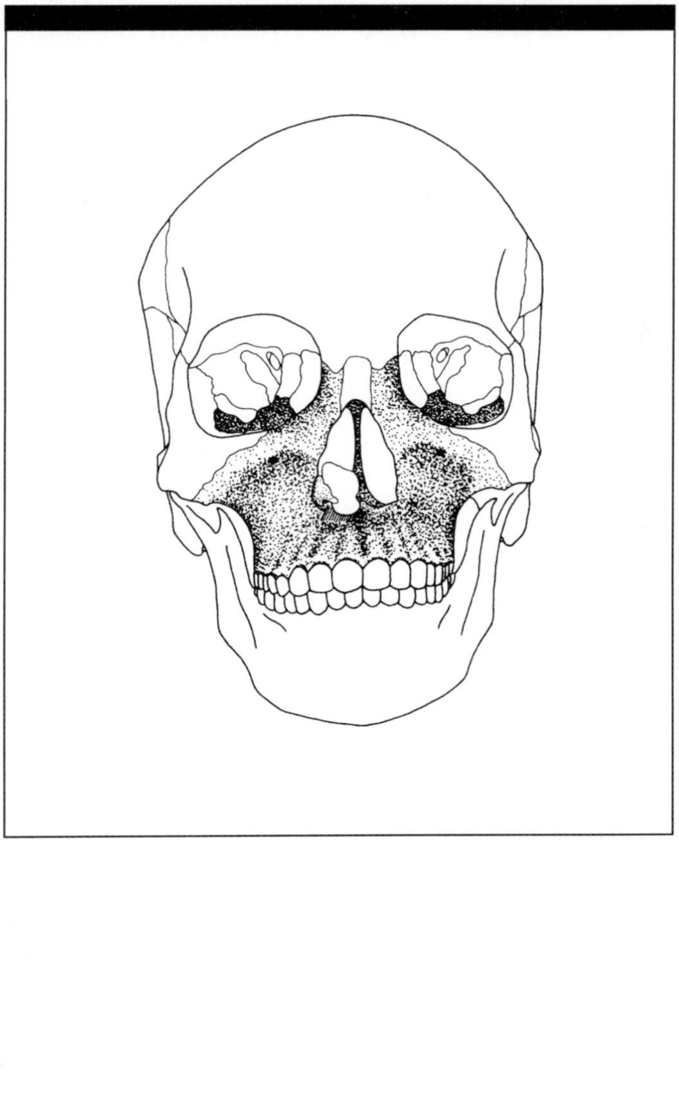

EPICRANIUS

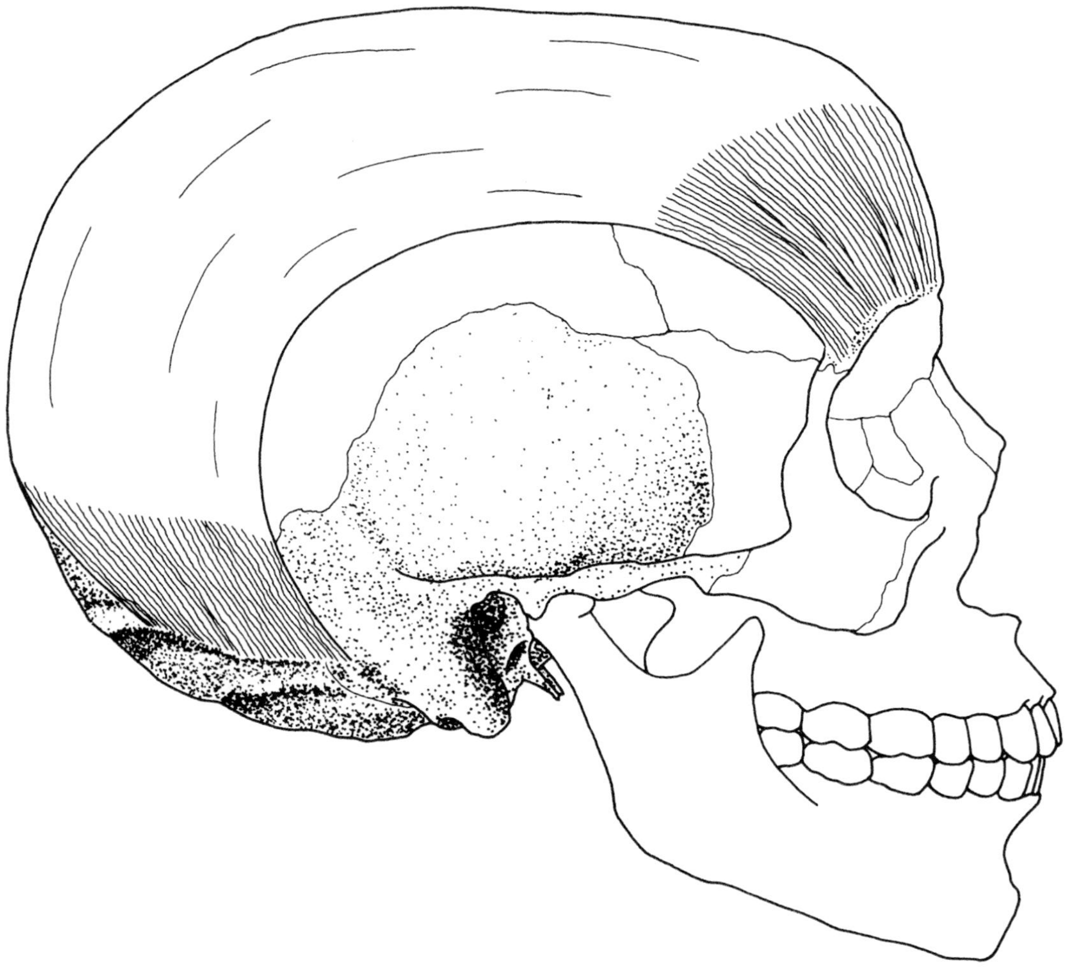

Skull—lateral view

Occipital belly *(occipitalis)*

Origin	Lateral two-thirds of superior nuchal line of occipital bone, mastoid process of temporal bone
Insertion	Galea aponeurotica (an intermediate tendon leading to frontal belly)
Action	Draws back scalp, aids frontal belly to wrinkle forehead and raise eyebrows
Nerve	Posterior auricular branch of facial nerve

Frontal belly *(frontalis)*

Origin	Galea aponeurotica
Insertion	Fascia of facial muscles and skin above nose and eyes
Action	Draws back scalp, wrinkles forehead, raises eyebrows
Nerve	Temporal branches of facial nerve

TEMPOROPARIETALIS

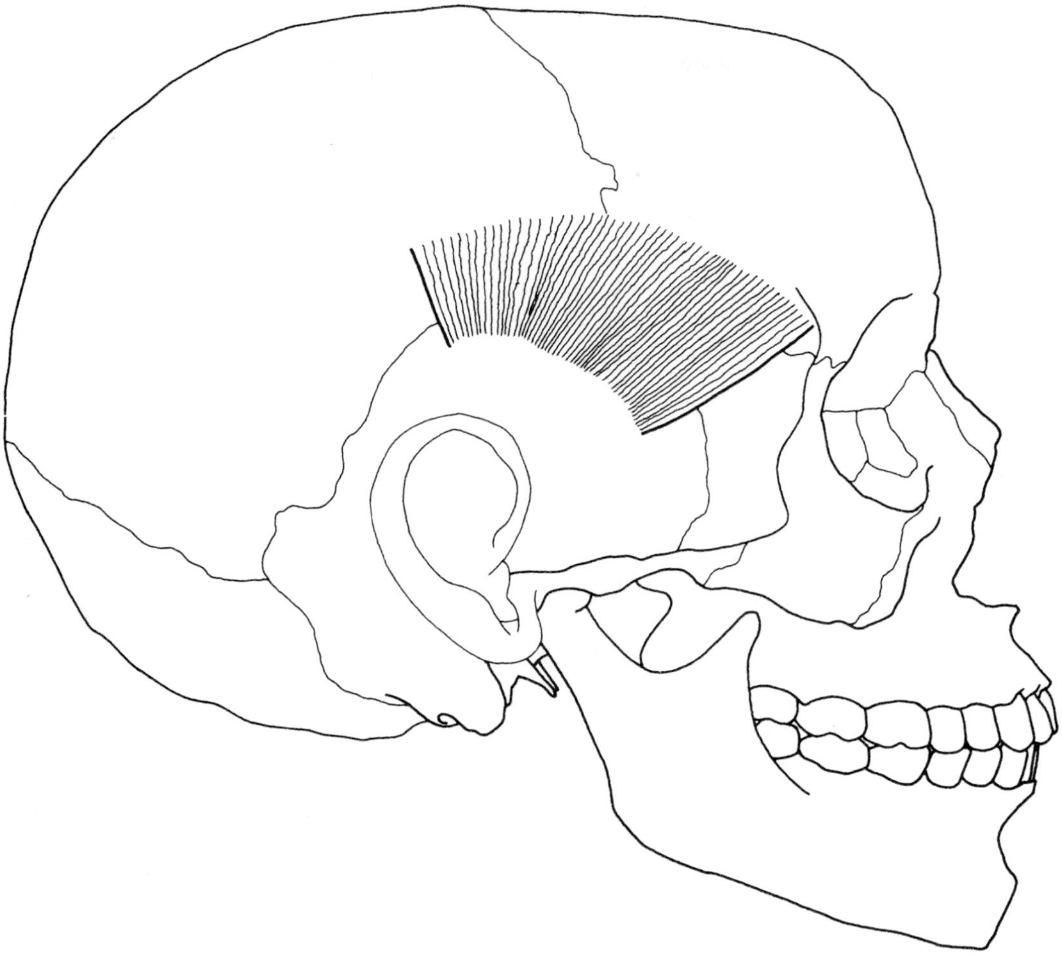

Skull—lateral view

Origin Fascia over ear

Insertion Lateral border of galea aponeurotica

Action Raises ears, tightens scalp

Nerve Temporal branch of facial nerve

AURICULARIS ANTERIOR, SUPERIOR, POSTERIOR

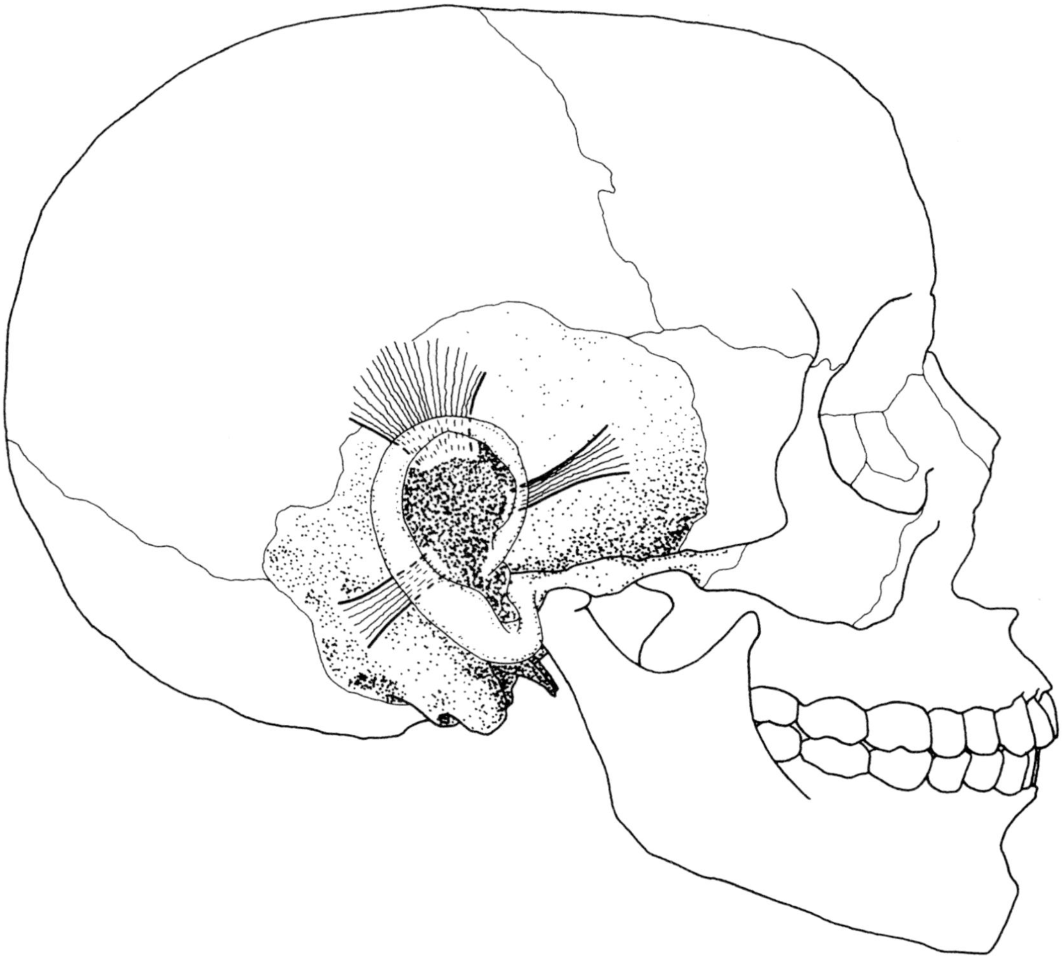

Skull—lateral view

Auricularis anterior

Origin	Fascia in temporal region
Insertion	Anterior to helix of ear
Action	Draws ear forward in some individuals, moves scalp*
Nerve	Temporal branch of facial nerve

Auricularis superior

Origin	Fascia in temporal region
Insertion	Superior part of ear
Action	Draws ear upward in some individuals, moves scalp*
Nerve	Temporal branch of facial nerve

Auricularis posterior

Origin	Mastoid area of temporal bone
Insertion	Posterior part of ear
Action	Draws ear upward in some individuals*
Nerve	Posterior auricular branch of facial nerve

*This muscle is nonfunctional in most people.

ORBICULARIS OCULI

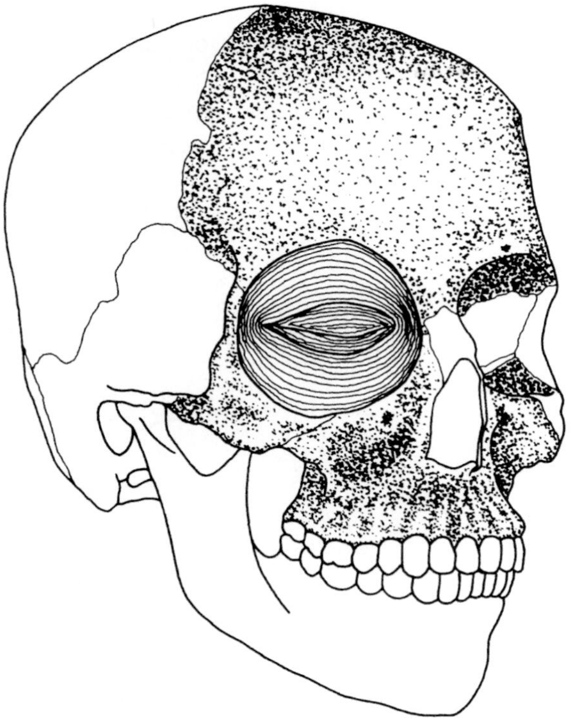

ORBITAL AND PALPEBRAL PARTS

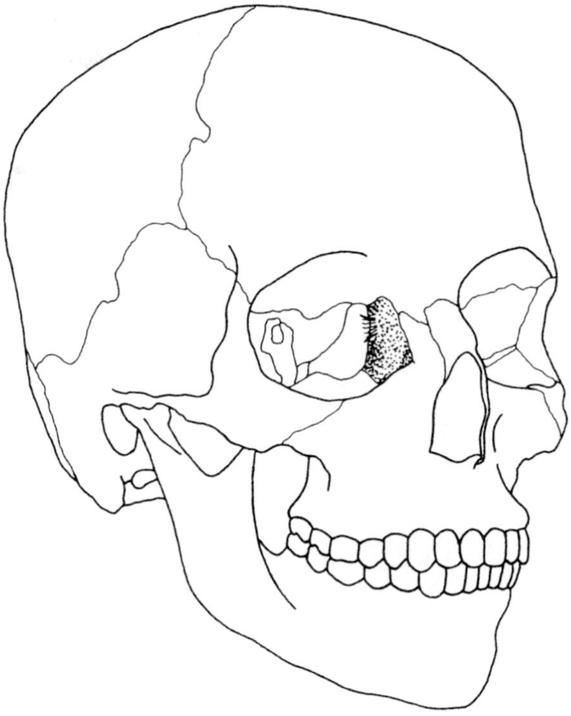

LACRIMAL PART

Skull—three-quarter anterior view

Orbital part

Origin	Frontal bone, maxilla (medial margin of orbit)
Insertion	Continues around orbit and returns to origin
Action	Strong closure of eyelids
Nerve	Temporal and zygomatic branches of facial nerve

Palpebral part *(in eyelids)*

Origin	Medial palpebral ligament
Insertion	Lateral palpebral ligament into zygomatic bone
Action	Gentle closure of eyelids
Nerve	Temporal and zygomatic branches of facial nerve

Lacrimal part *(behind medial palpebral ligament and lacrimal sac)*

Origin	Lacrimal bone
Insertion	Lateral palpebral raphe
Action	Draws lacrimal canals onto surface of eye
Nerve	Temporal and zygomatic branches of facial nerve

LEVATOR PALPEBRAE SUPERIORIS

Origin	Inferior surface of lesser wing of sphenoid
Insertion	Skin of upper eyelid
Action	Raises upper eyelid
Nerve	Oculomotor nerve

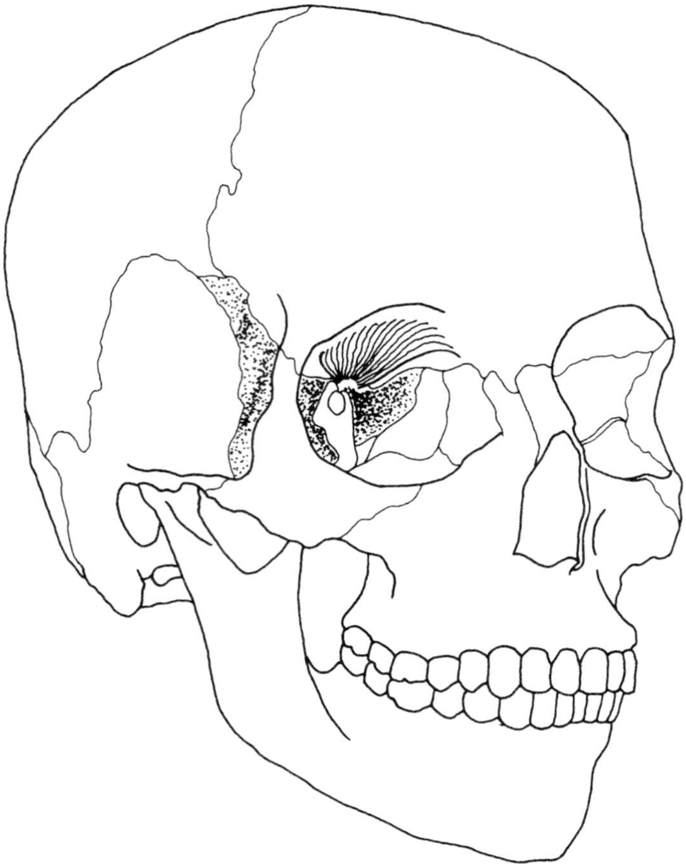

Skull—three-quarter anterior view

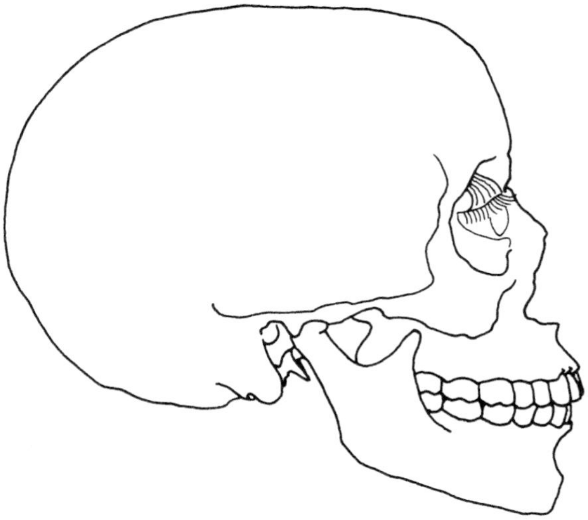

Skull—lateral view

CORRUGATOR SUPERCILII

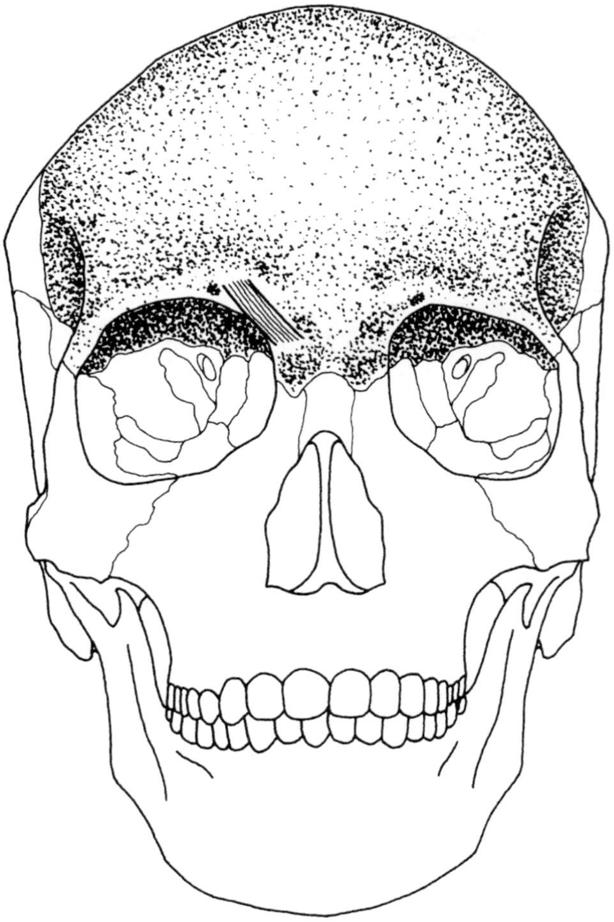

Skull—anterior view

Origin Medial end of superciliary arch

Insertion Deep surface of skin under medial portion of eyebrows

Action Draws eyebrows downward and medially

Nerve Temporal branch of facial nerve

PROCERUS

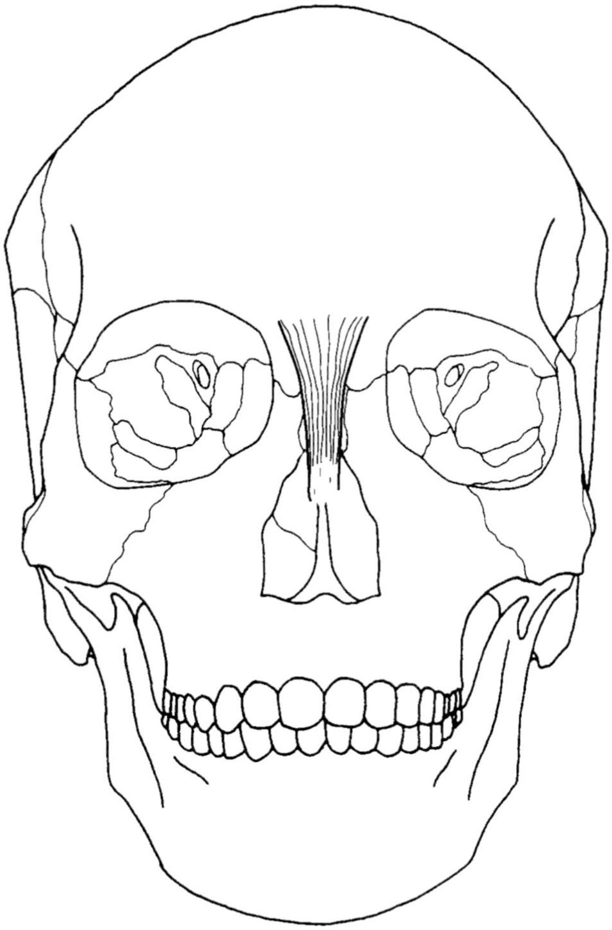

Skull—anterior view

Origin Fascia over nasal bone and lateral nasal cartilage

Insertion Skin between eyebrows

Action Draws down medial part of eyebrows, wrinkles nose

Nerve Buccal branches of facial nerve

NASALIS

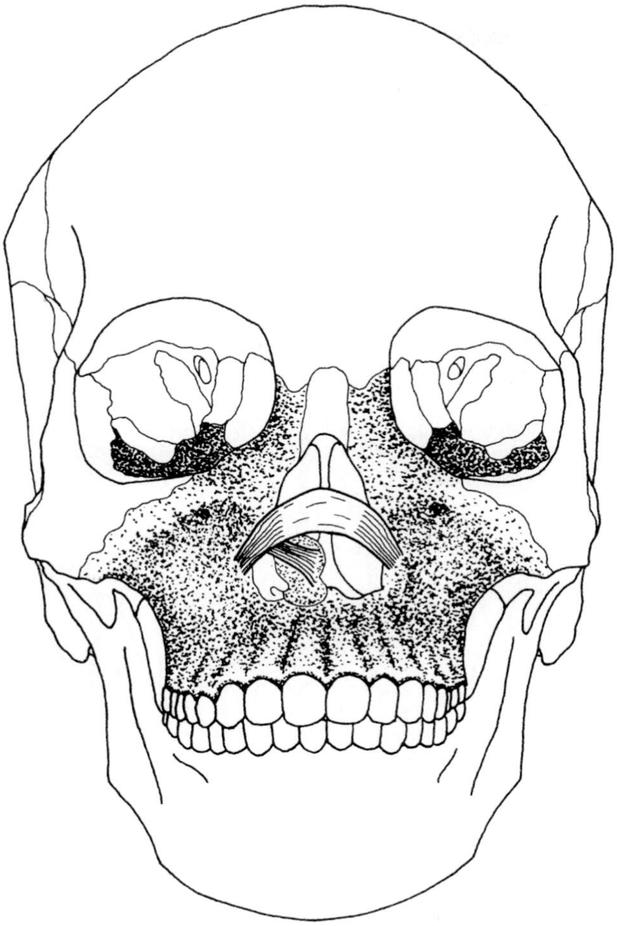

Skull—anterior view

Transverse part

Origin	Middle of maxilla
Insertion	Muscle of opposite side over bridge of nose

Alar part

Origin	Greater alar cartilage, skin on nose
Insertion	Skin at point of nose
Action	Both parts maintain opening of external nares during forceful inspiration
Nerve	Buccal branches of facial nerve

DEPRESSOR SEPTI

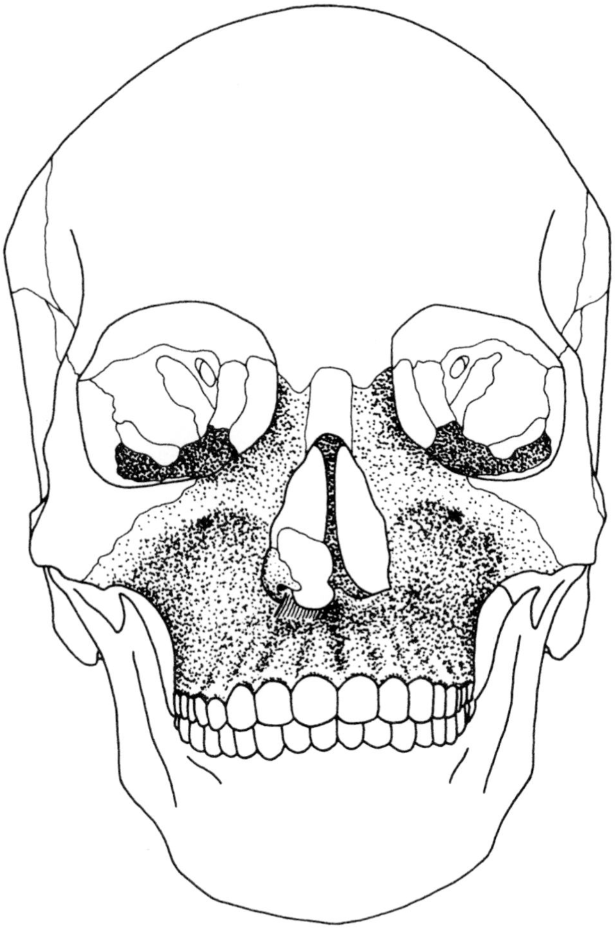

Skull—anterior view

Origin	Incisive fossa of maxilla	**Action**	Constricts nares
Insertion	Nasal septum and ala	**Nerve**	Buccal branches of facial nerve

ORBICULARIS ORIS

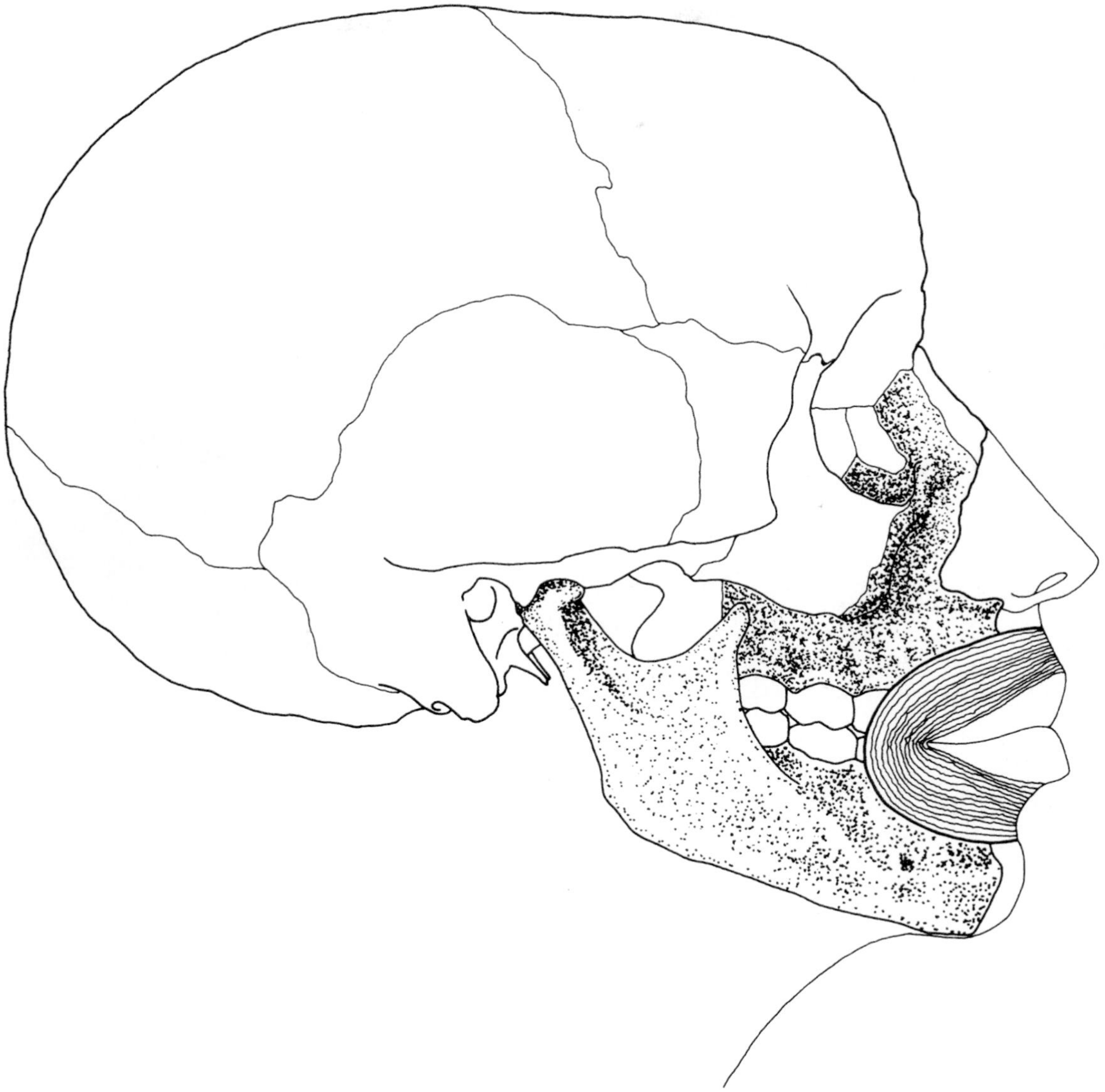

Skull—lateral view

Origin	Lateral band—alveolar border of maxilla Medial band—septum of nose Inferior portion—lateral to midline of mandible	**Insertion**	Becomes continuous with other muscles at angle of mouth
		Action	Closure and protrusion of lips
		Nerve	Buccal and mandibular branches of facial nerve

LEVATOR LABII SUPERIORIS

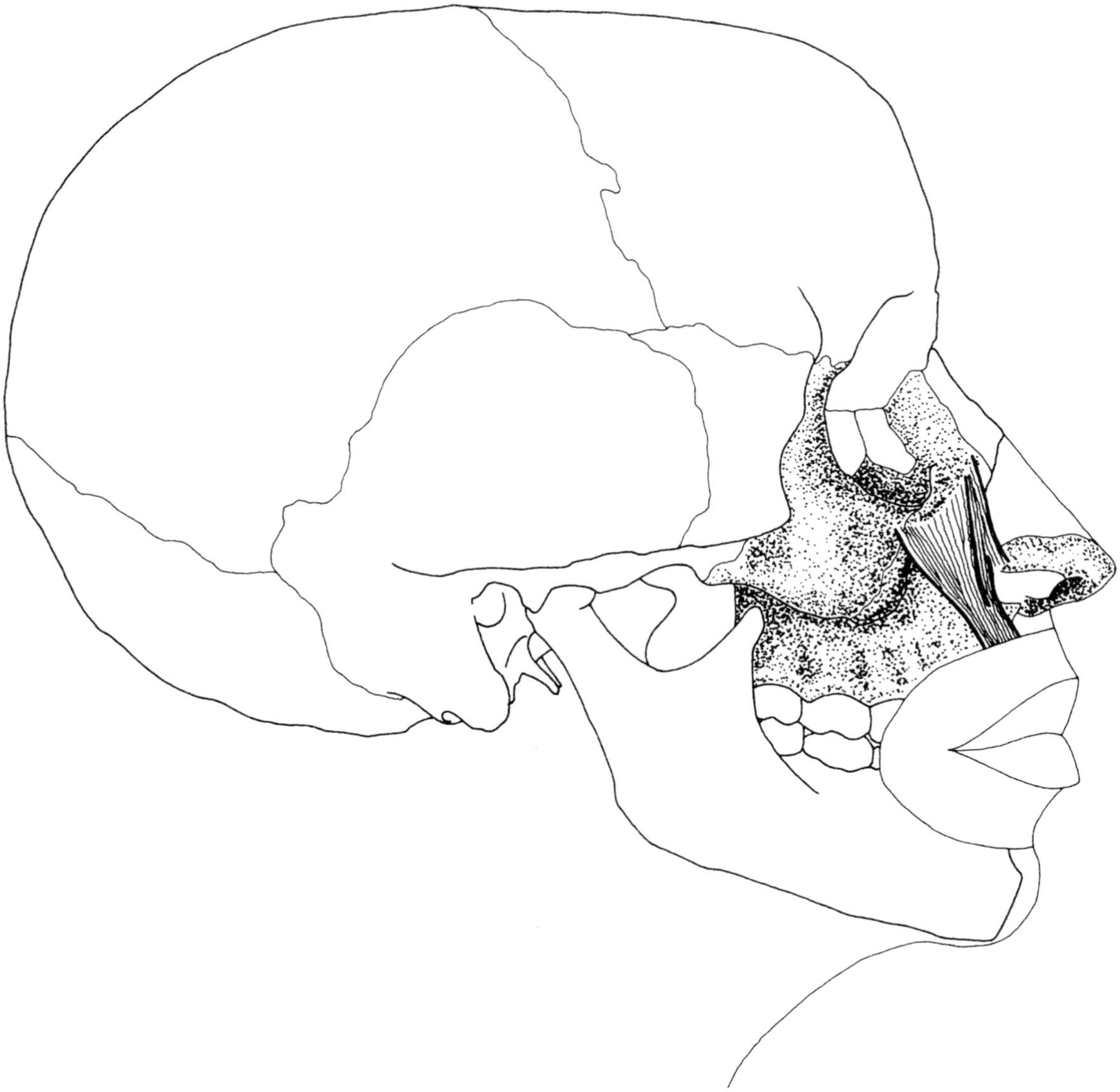

Skull—lateral view

Angular head

Origin	Frontal process of maxilla and zygomatic bone
Insertion	Greater alar cartilage and skin of nose, upper lip
Action	Elevates upper lip, dilates nares, forms nasolabial furrow
Nerve	Buccal branches of facial nerve

Infraorbital head

Origin	Lower margin of orbit
Insertion	Muscles of upper lip
Action	Elevates upper lip
Nerve	Buccal branches of facial nerve

LEVATOR ANGULI ORIS

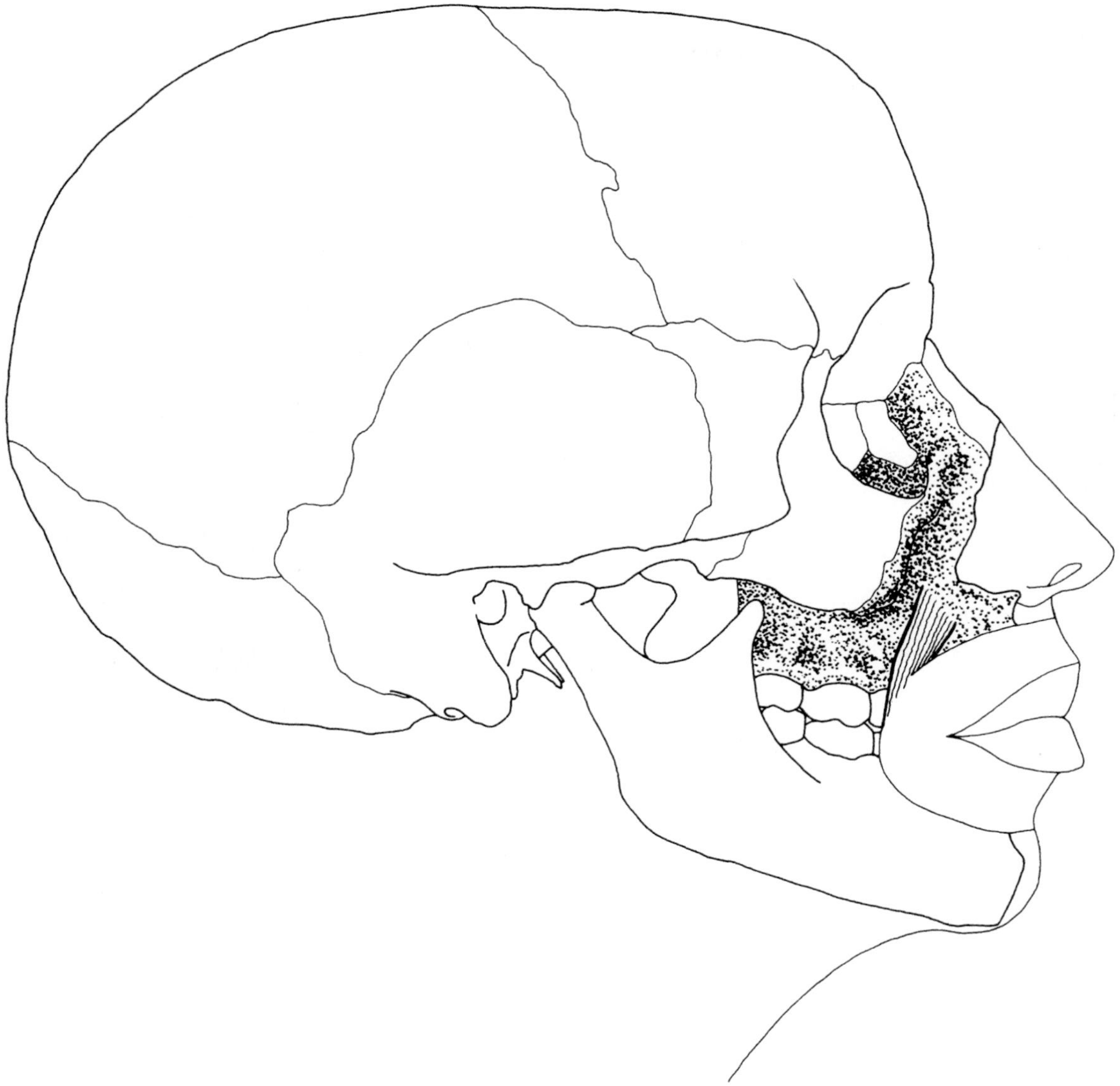

Skull—lateral view

Origin	Canine fossa of maxilla	**Action**	Elevates corner (angle) of mouth
Insertion	Angle of mouth	**Nerve**	Buccal branches of facial nerve

ZYGOMATICUS MAJOR

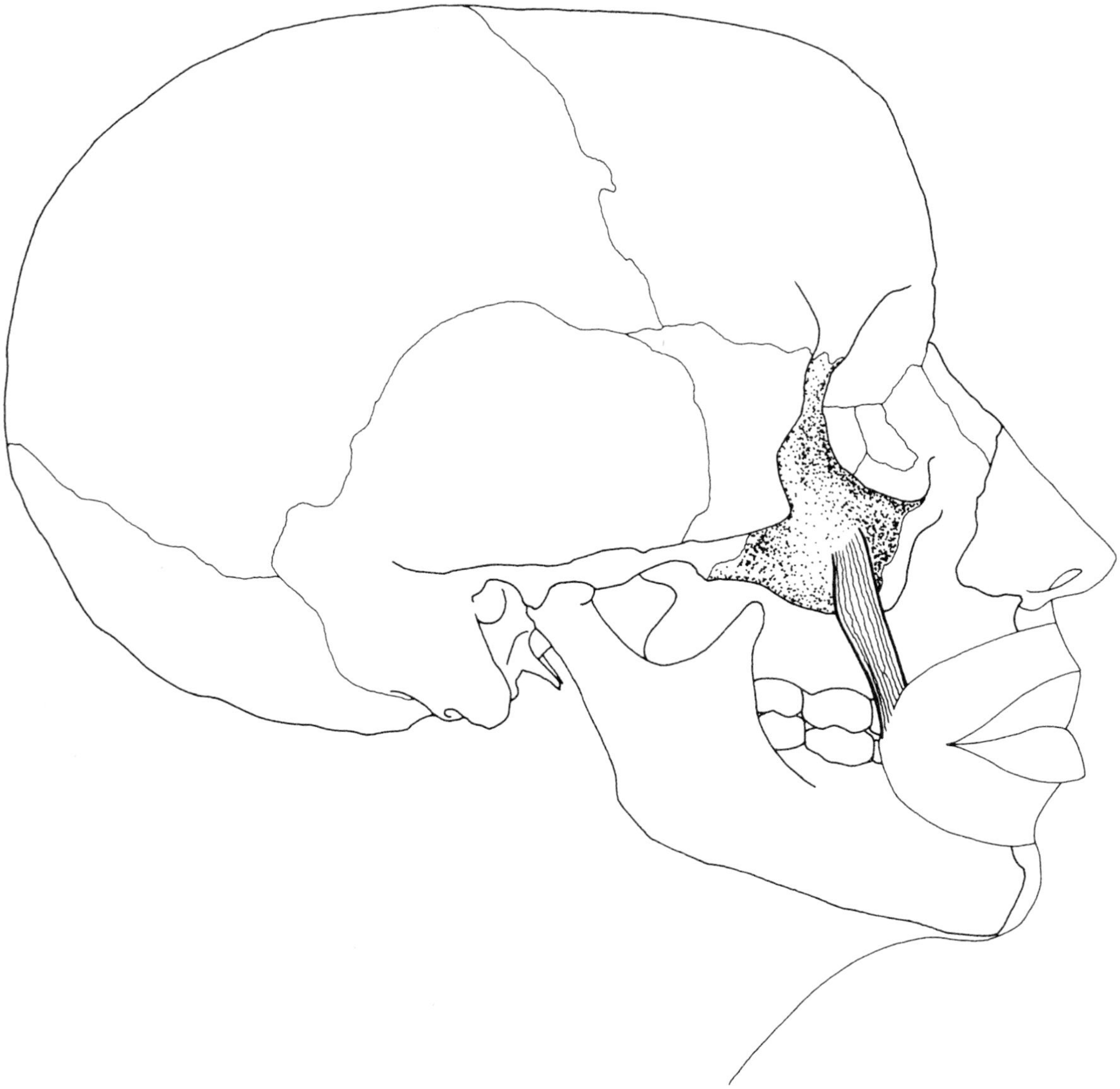

Skull—lateral view

Origin Zygomatic bone

Insertion Angle of mouth

Action Draws angle of mouth upward and backward (laughing)

Nerve Buccal branches of facial nerve

ZYGOMATICUS MINOR

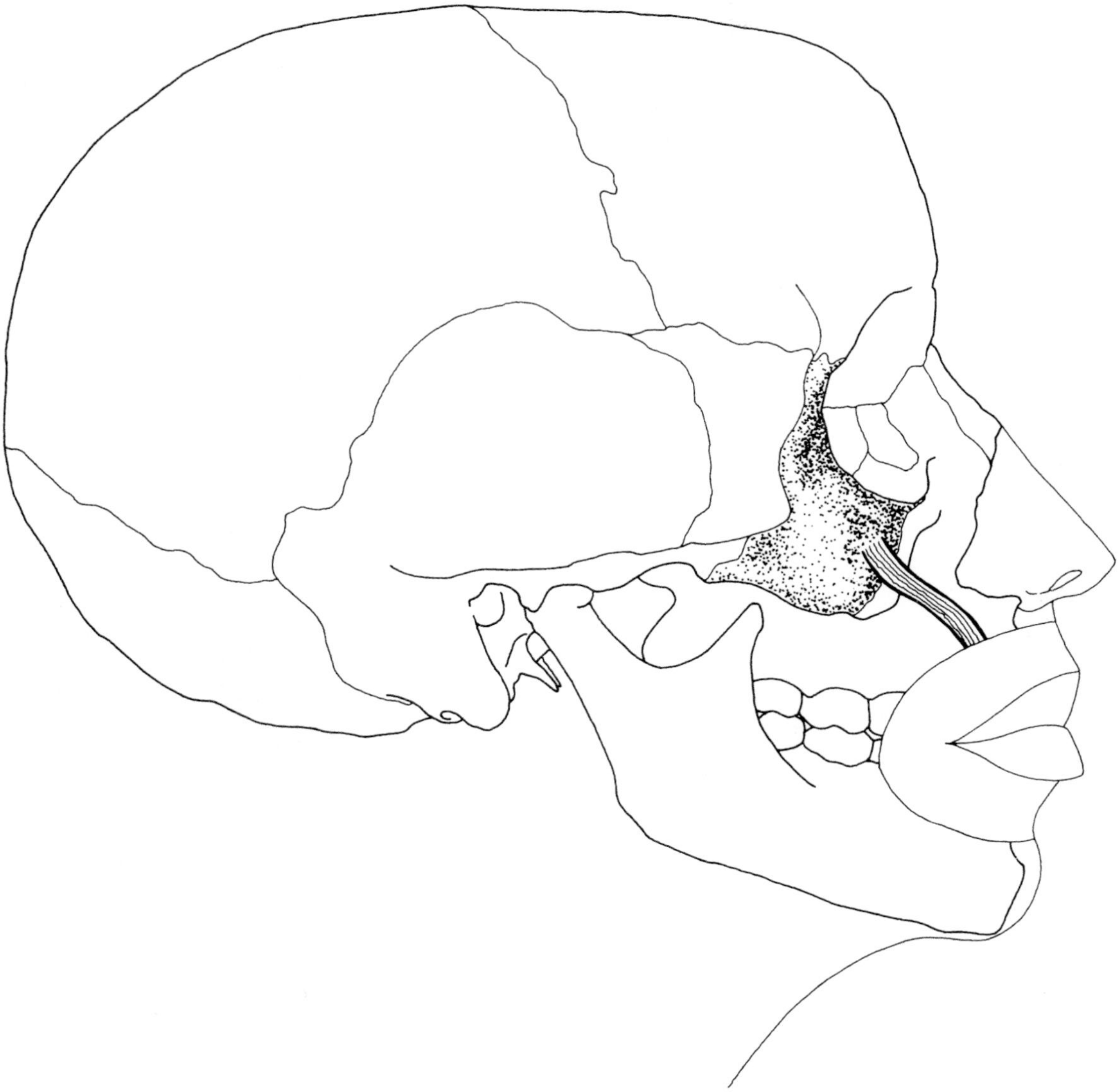

Skull—lateral view

Origin Zygomatic bone

Insertion Upper lip lateral to levator labii superioris

Action Forms nasolabial furrow

Nerve Buccal branches of facial nerve

RISORIUS

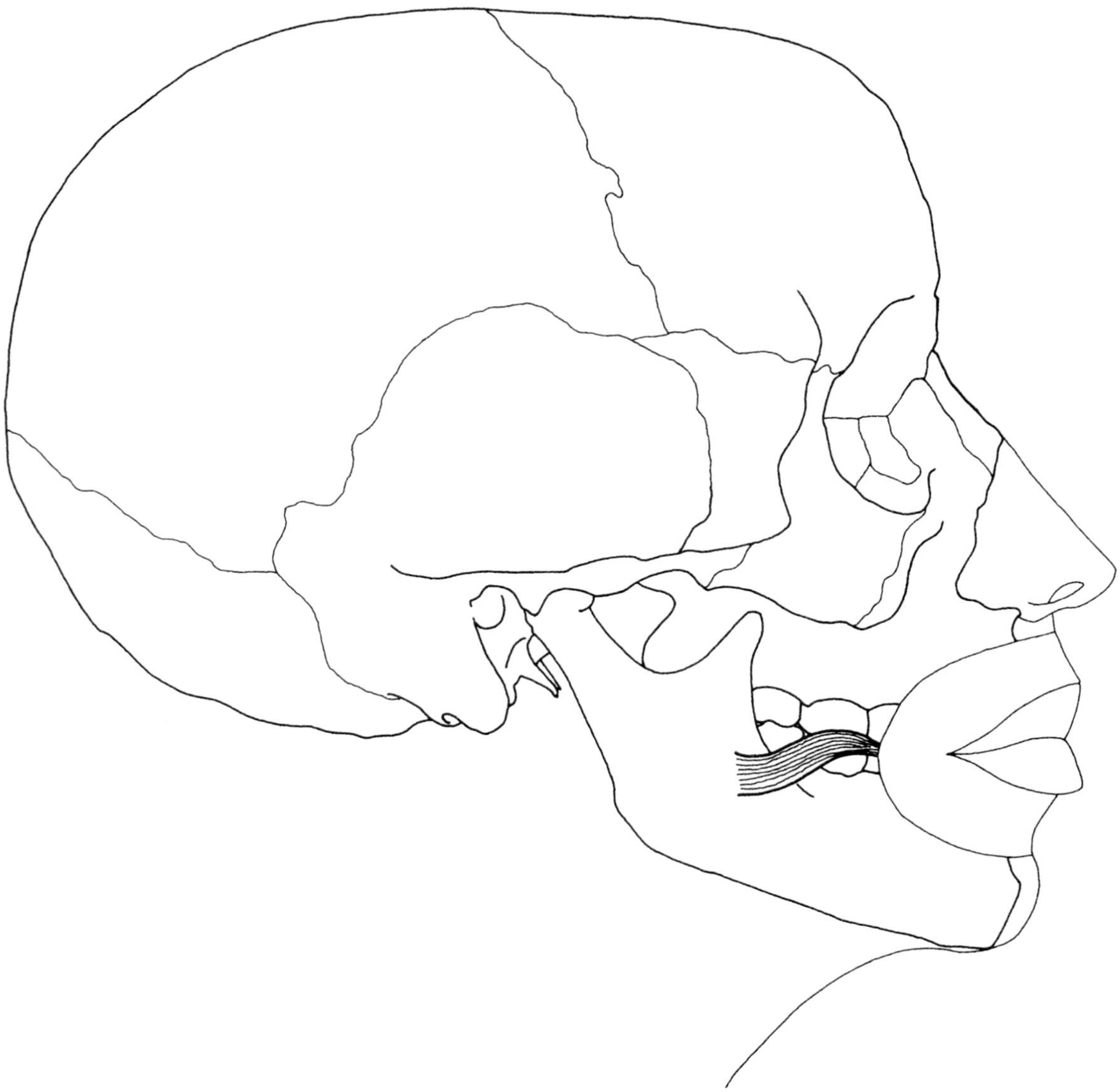

Skull—lateral view

Origin Fascia over masseter

Insertion Skin at angle of mouth

Action Retracts angle of mouth, as in grinning

Nerve Buccal branches of facial nerve

DEPRESSOR LABII INFERIORIS

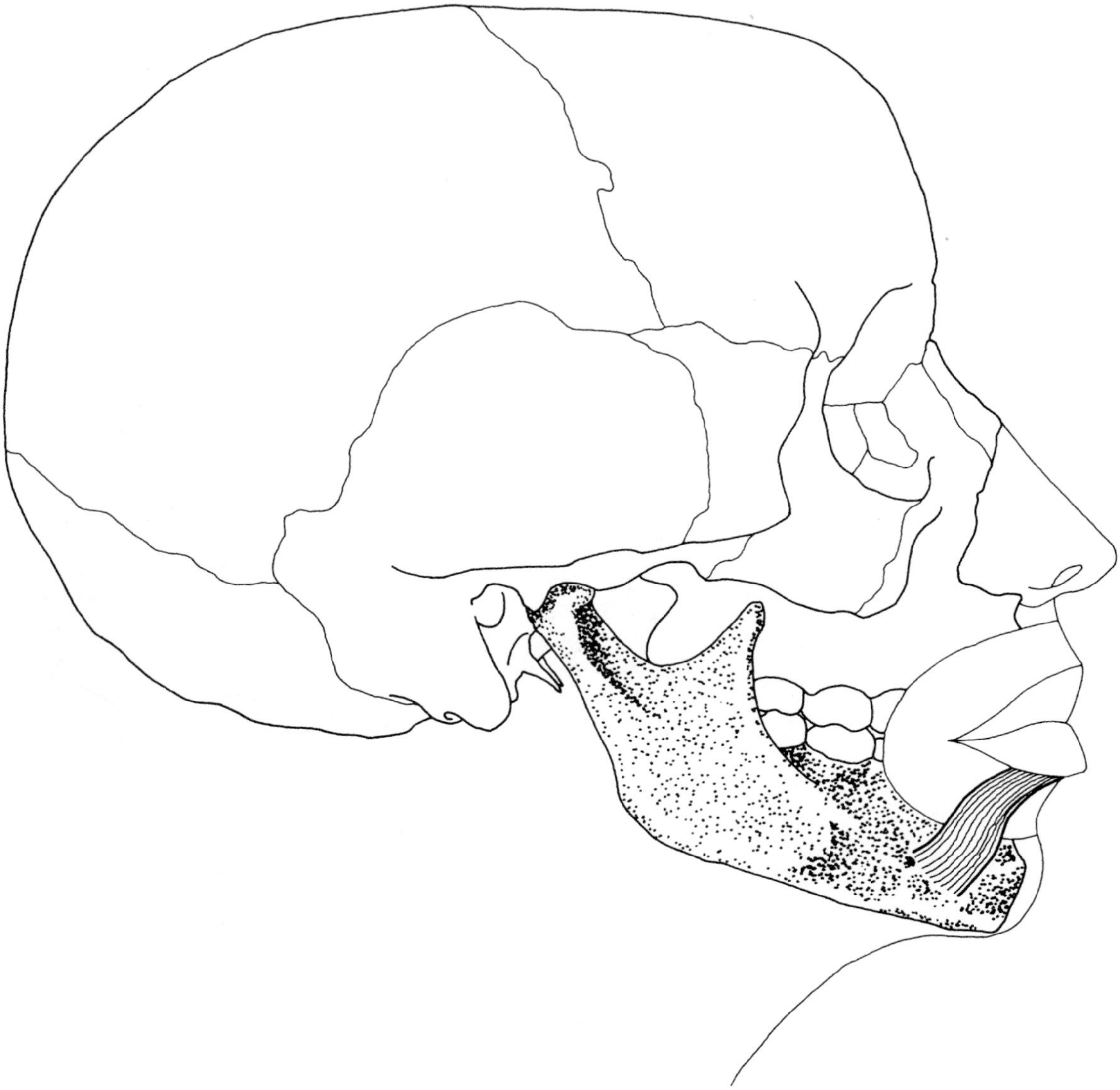

Skull—lateral view

Origin Mandible, between symphysis and mental foramen

Insertion Skin of lower lip

Action Draws lower lip downward and laterally

Nerve Mandibular branch of facial nerve

DEPRESSOR ANGULI ORIS

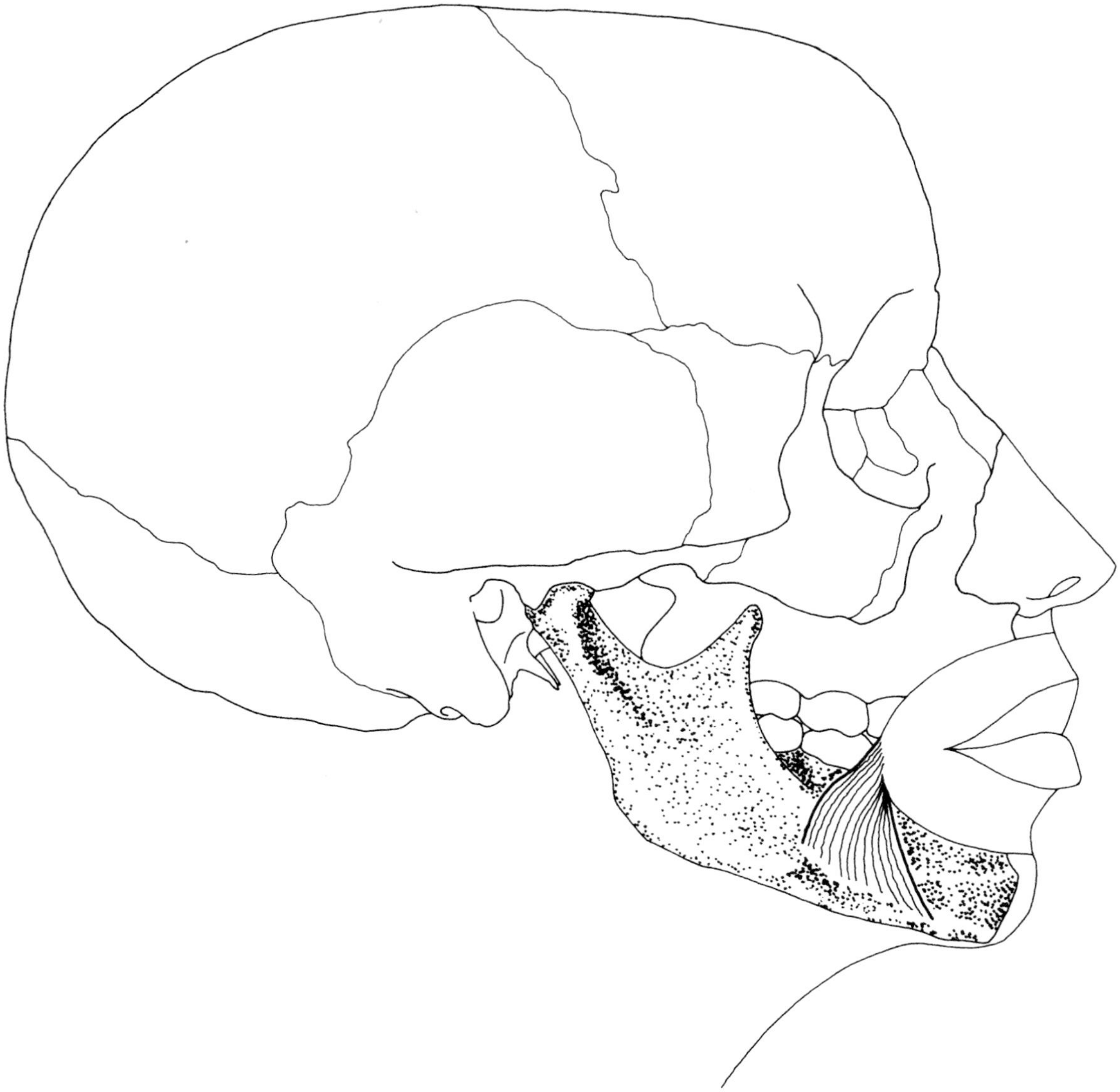

Skull—lateral view

Origin Oblique line of the mandible
Insertion Angle of the mouth

Action Depresses angle of mouth, as in frowning
Nerve Mandibular branch of facial nerve

MENTALIS

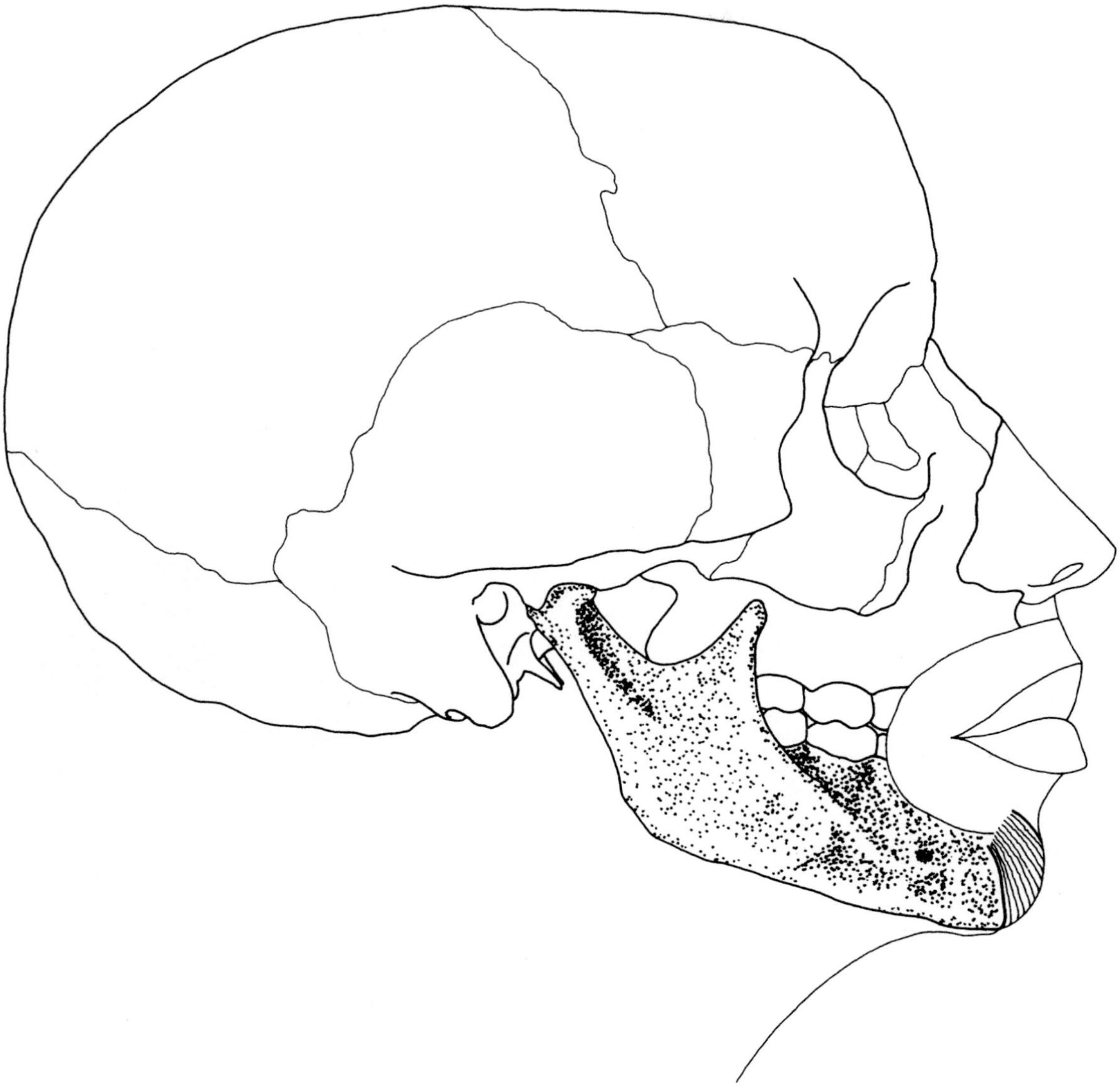

Skull—lateral view

Origin Incisive fossa of mandible

Insertion Skin of chin

Action Raises and protrudes lower lip, wrinkles skin of chin

Nerve Mandibular branch of facial nerve

BUCCINATOR

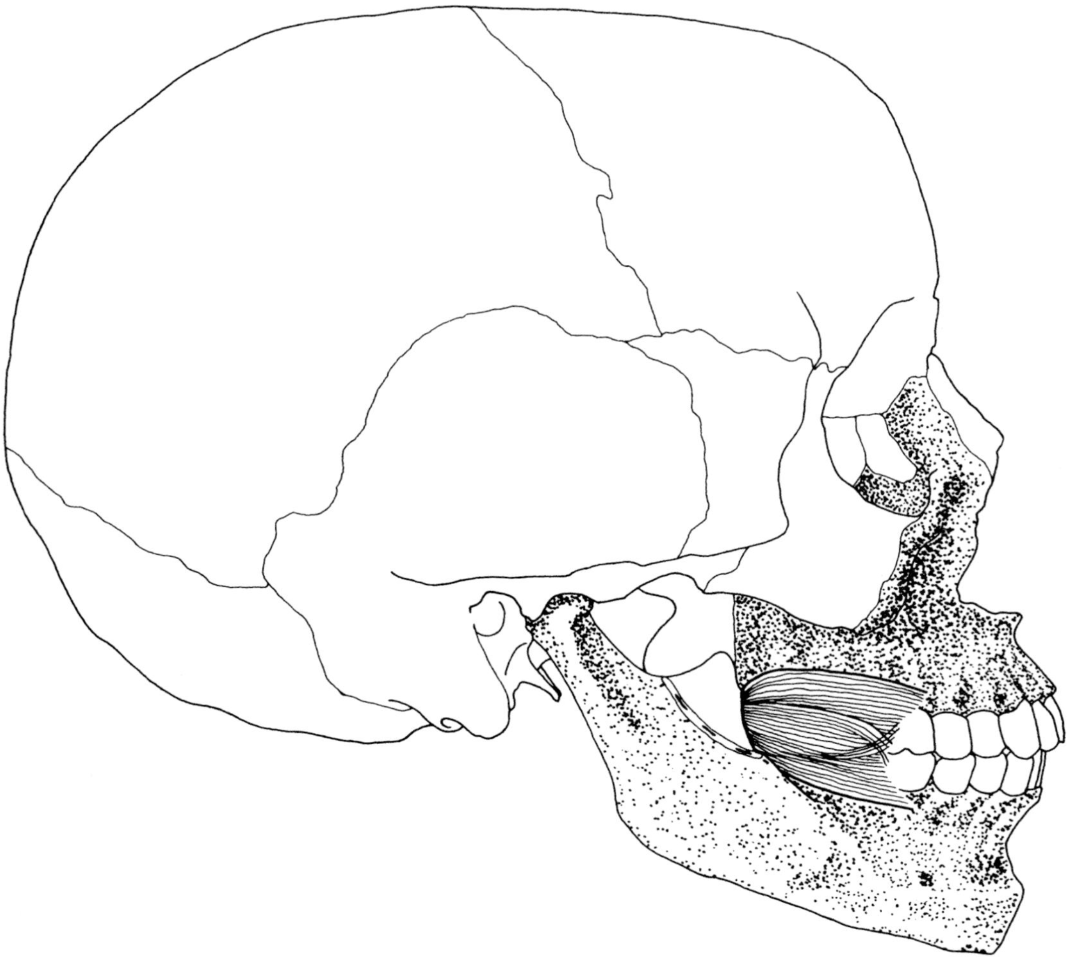

Skull—lateral view

Origin Outer surface of alveolar processes of maxilla and mandible over molars and along pterygomandibular raphe

Insertion Deep part of muscles of lips

Action Compresses cheek

Nerve Buccal branches of facial nerve

TEMPORALIS

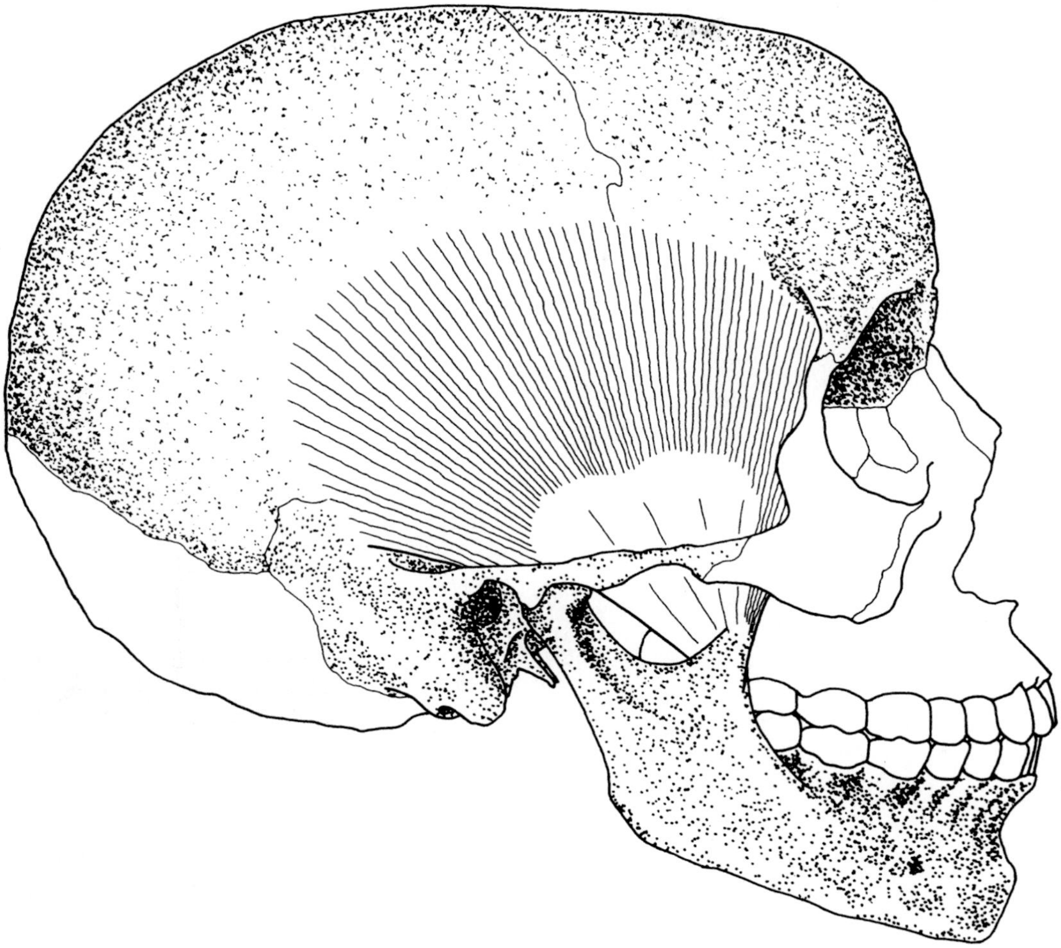

Skull—lateral view

Origin Temporal fossa including frontal, parietal, and temporal bones

Insertion Coronoid process and anterior border of ramus of mandible

Action Closes lower jaw, clenches teeth

Nerve Mandibular division of trigeminal nerve

MASSETER

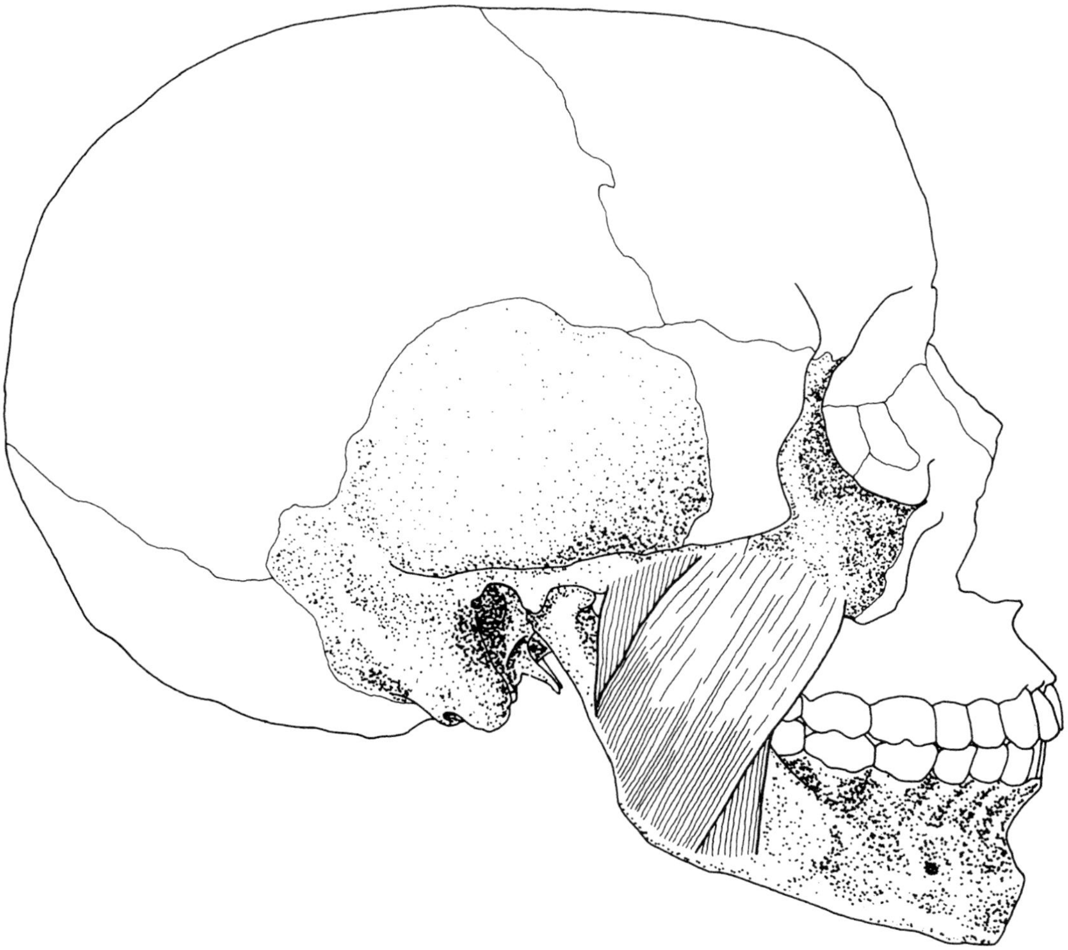

Skull—lateral view

Origin Zygomatic process of maxilla, medial and inferior surfaces of zygomatic arch

Insertion Angle and ramus of mandible, lateral surface of coronoid process of mandible

Action Closes lower jaw, clenches teeth

Nerve Mandibular division of trigeminal nerve

Note: Superficial fibers slightly protract jaw (see lateral pterygoid).

PTERYGOIDEUS MEDIALIS

(Medial Pterygoid)

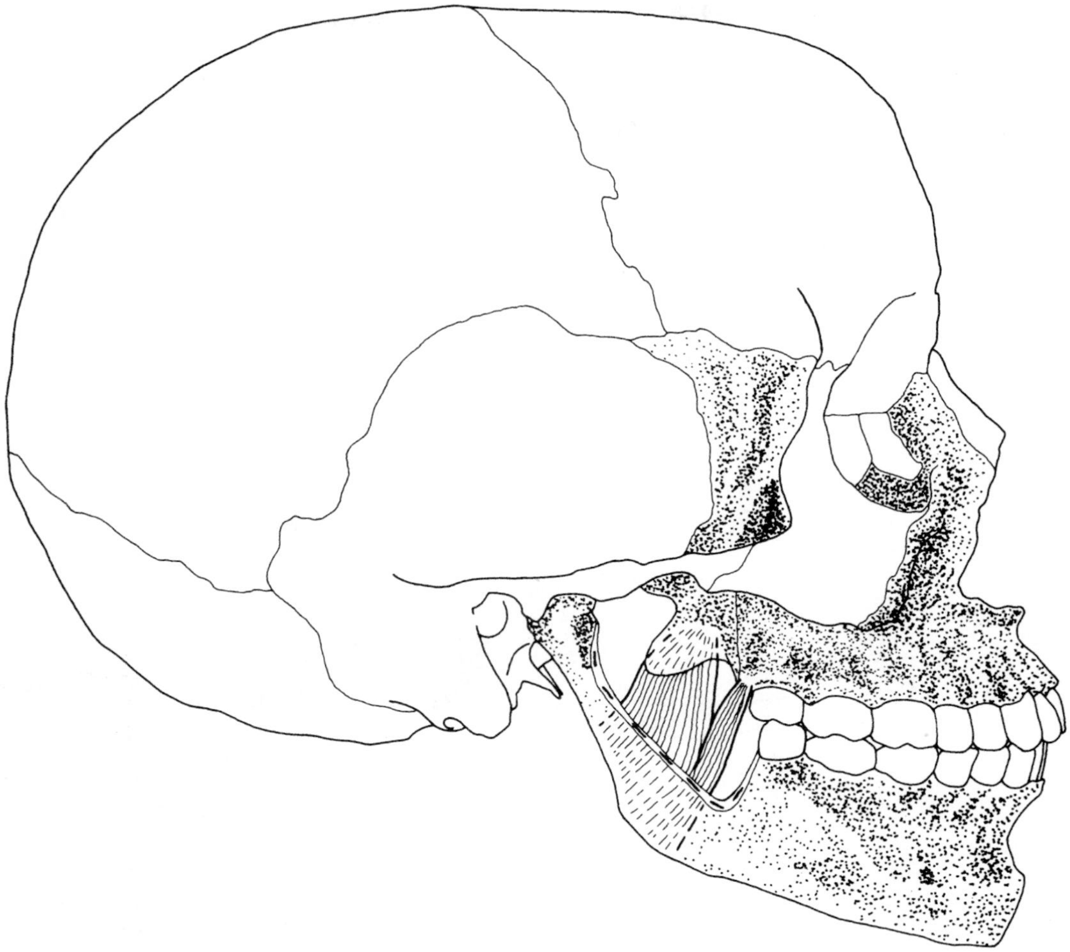

Skull—lateral view

(Part of mandible cut away)

Origin Medial surface of lateral pterygoid plate of sphenoid bone, palatine bone, and tuberosity of maxilla

Insertion Medial surface of ramus and angle of mandible

Action Closes lower jaw, clenches teeth

Nerve Mandibular division of trigeminal nerve

PTERYGOIDEUS LATERALIS*

(Lateral Pterygoid)

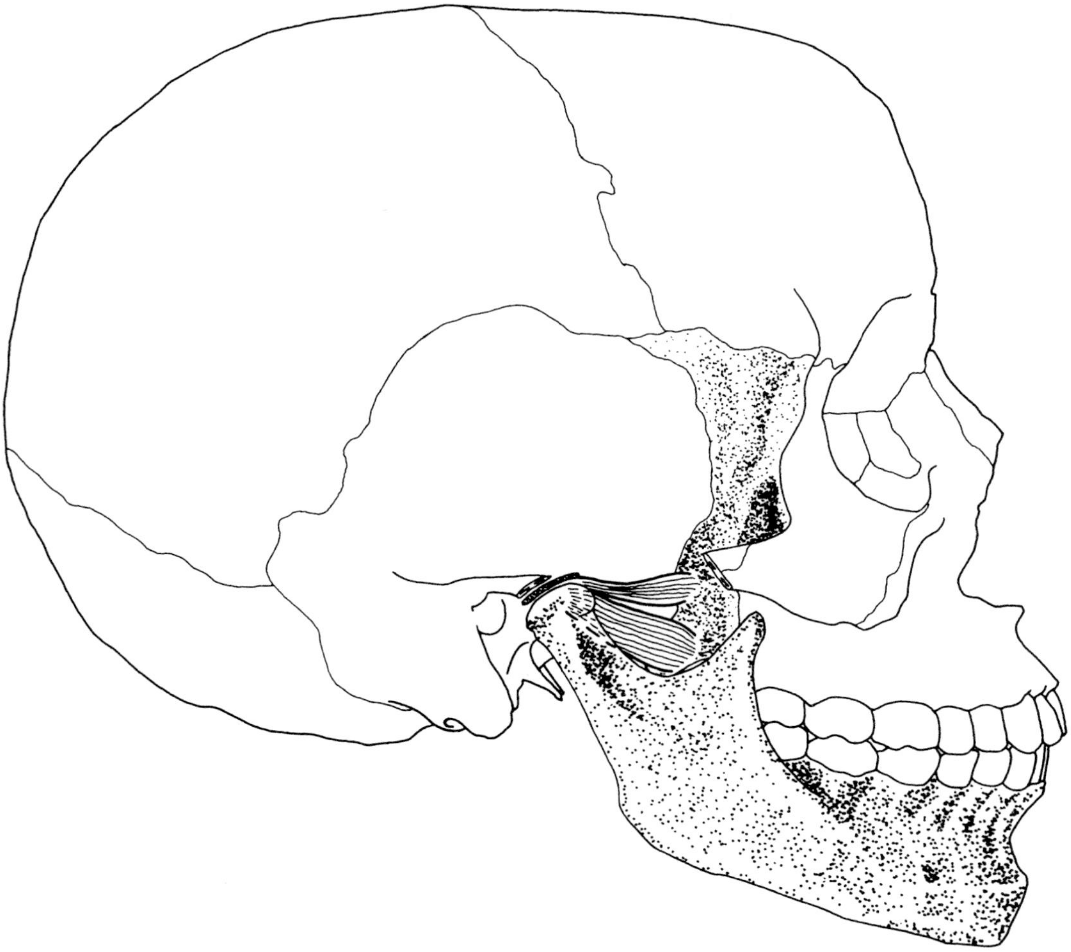

Skull—lateral view

Origin Superior head*—lateral surface of greater wing of sphenoid

Inferior head—lateral surface of lateral pterygoid plate

Insertion Condyle of mandible, temporomandibular joint

Action Opens jaws, protrudes mandible, moves mandible sidewards

Nerve Mandibular division of trigeminal nerve

Note: This sideward movement, aided by superficial fibers of masseter, causes chewing movements.

*Stern calls this a separate muscle: superior pterygoid.

Reference: Stern, JT: *Essentials of Gross Anatomy,* F. A. Davis Company, Philadelphia, 1988.

CHAPTER FOUR
MUSCLES OF THE NECK

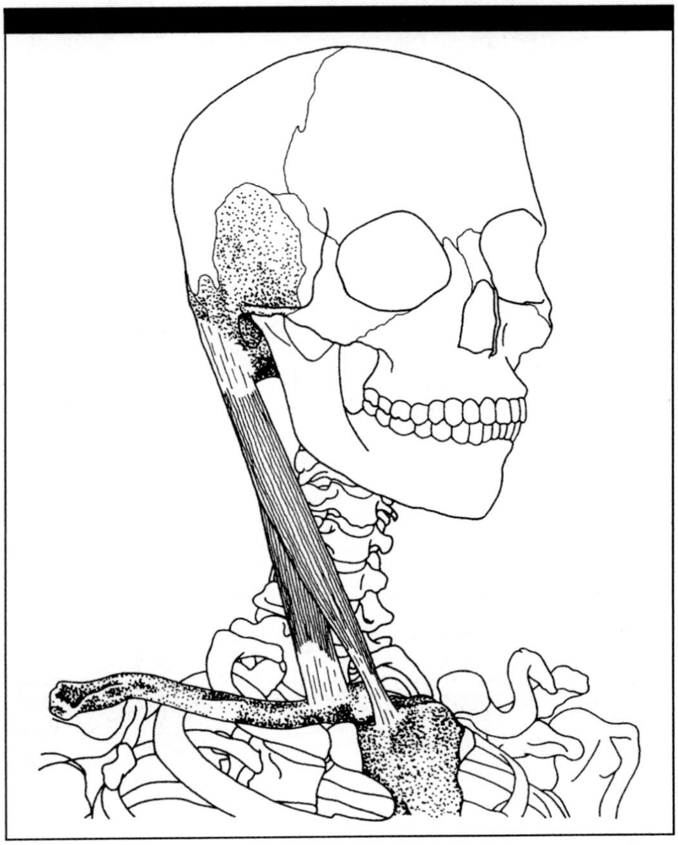

STERNOCLEIDOMASTOIDEUS

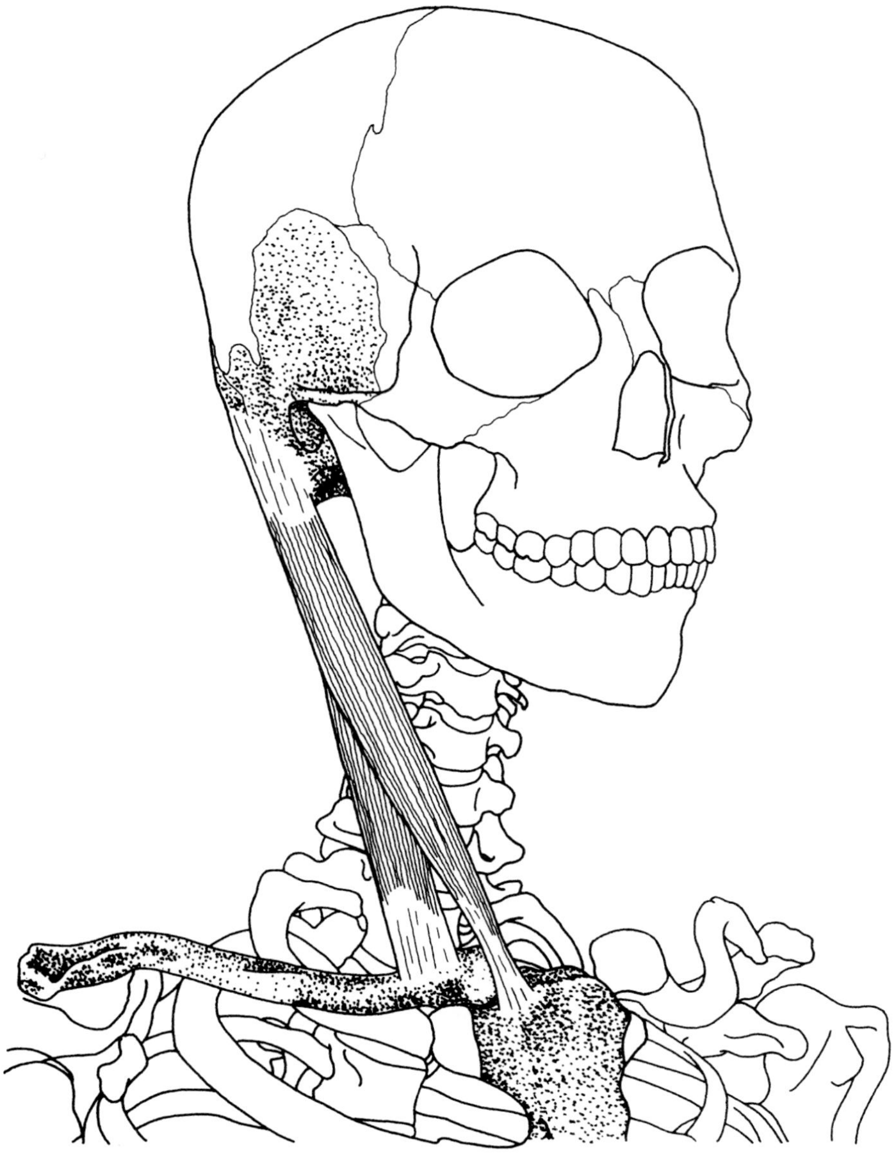

Three-quarter frontal view

Origin Sternal head—manubrium of sternum
Clavicular head—medial part of clavicle

Insertion Mastoid process of temporal bone, lateral half of superior nuchal line of occipital bone

Action One side—bends neck laterally, rotates head to opposite side
Both sides together—flexes neck, draws head ventrally and elevates chin, draws sternum superiorly in deep inspiration

Nerve Spinal part of accessory nerve (C2, C3)

PLATYSMA

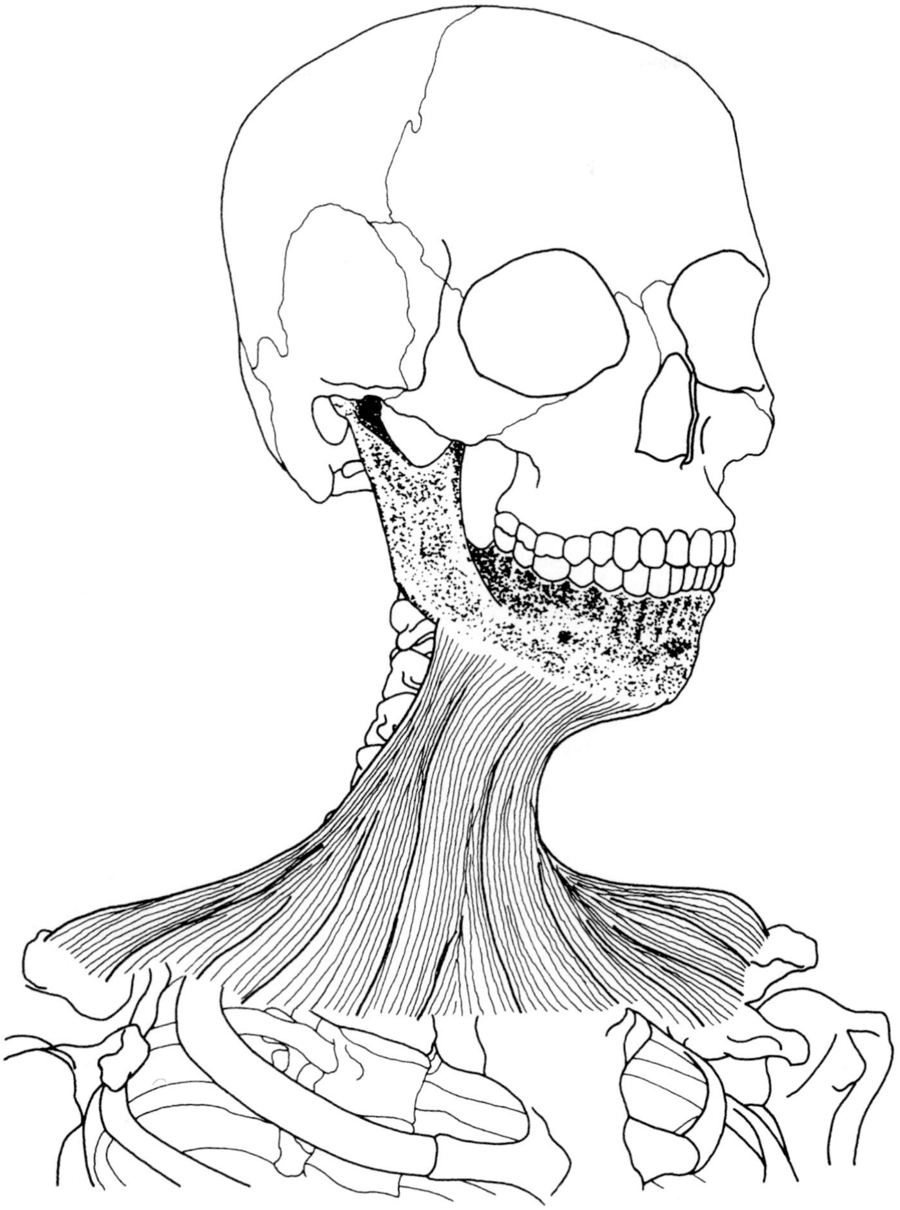

Three-quarter frontal view

Origin	Subcutaneous fascia of upper one-fourth of chest	**Action**	Depresses and draws lower lip laterally, draws up skin of chest
Insertion	Subcutaneous fascia and muscles of chin and jaw, mandible	**Nerve**	Cervical branch of facial nerve

DIGASTRICUS

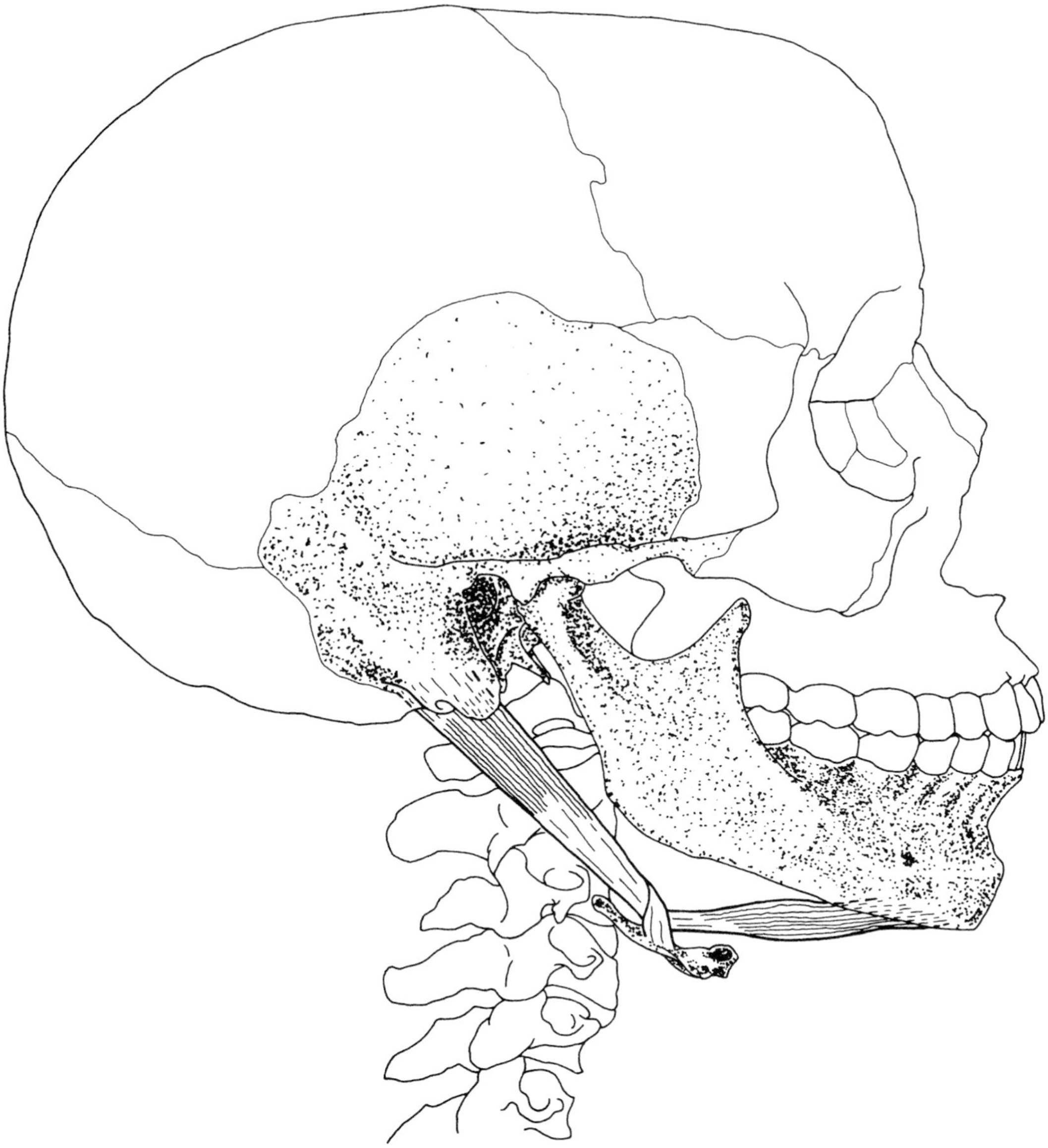

Lateral view

Origin

Posterior belly—mastoid notch of temporal bone

Anterior belly—inner side of inferior border of mandible near symphysis

Insertion

Intermediate tendon attached to hyoid bone

Action

Raises hyoid bone, assists in opening jaws, moves hyoid forward or backward

Nerve

Anterior belly—mandibular division of trigeminal

Posterior belly—facial nerve

STYLOHYOIDEUS

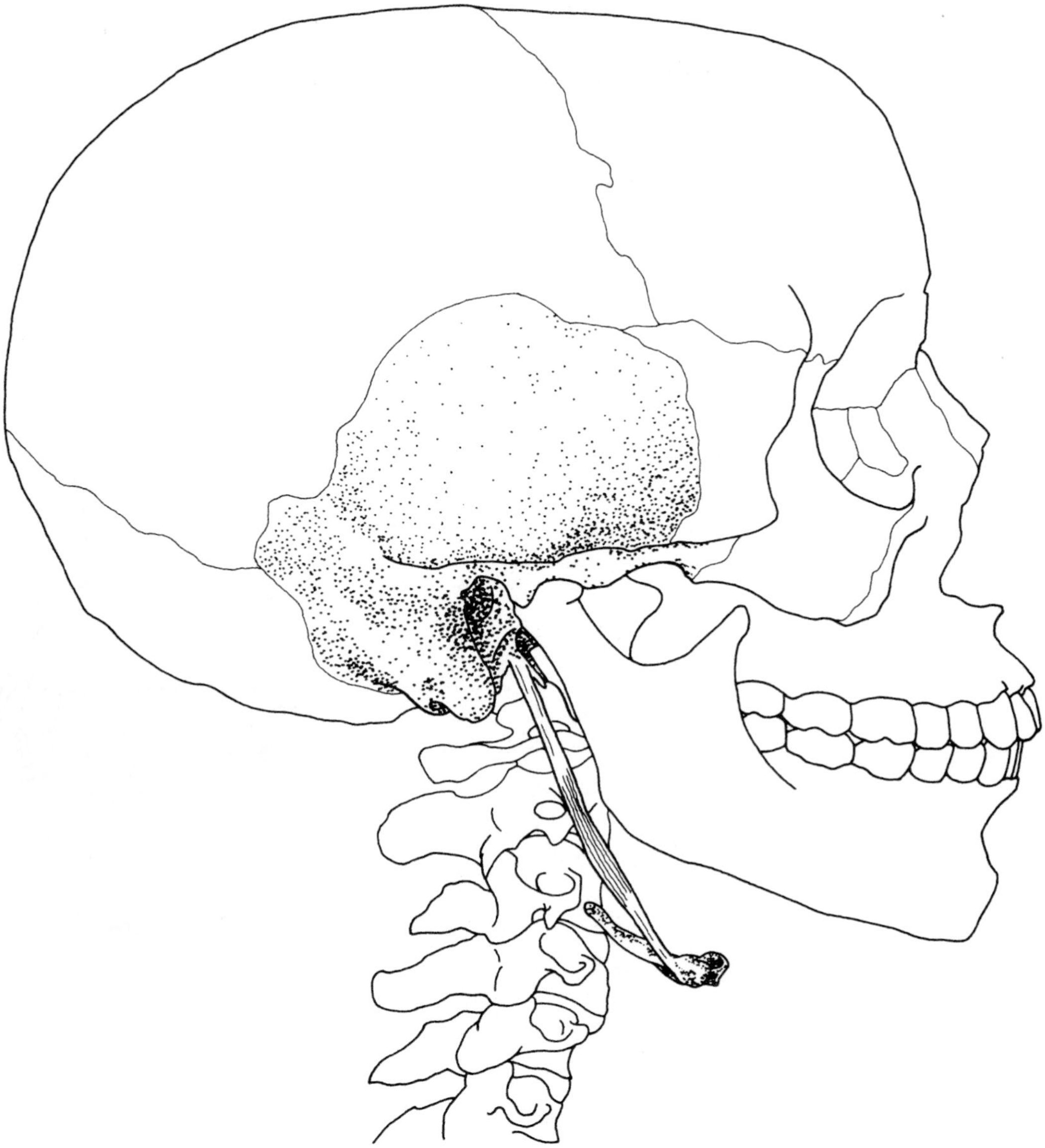

Lateral view

Origin Styloid process of temporal bone
Insertion Hyoid bone

Action Draws hyoid bone backward, elevates tongue
Nerve Facial nerve

MYLOHYOIDEUS

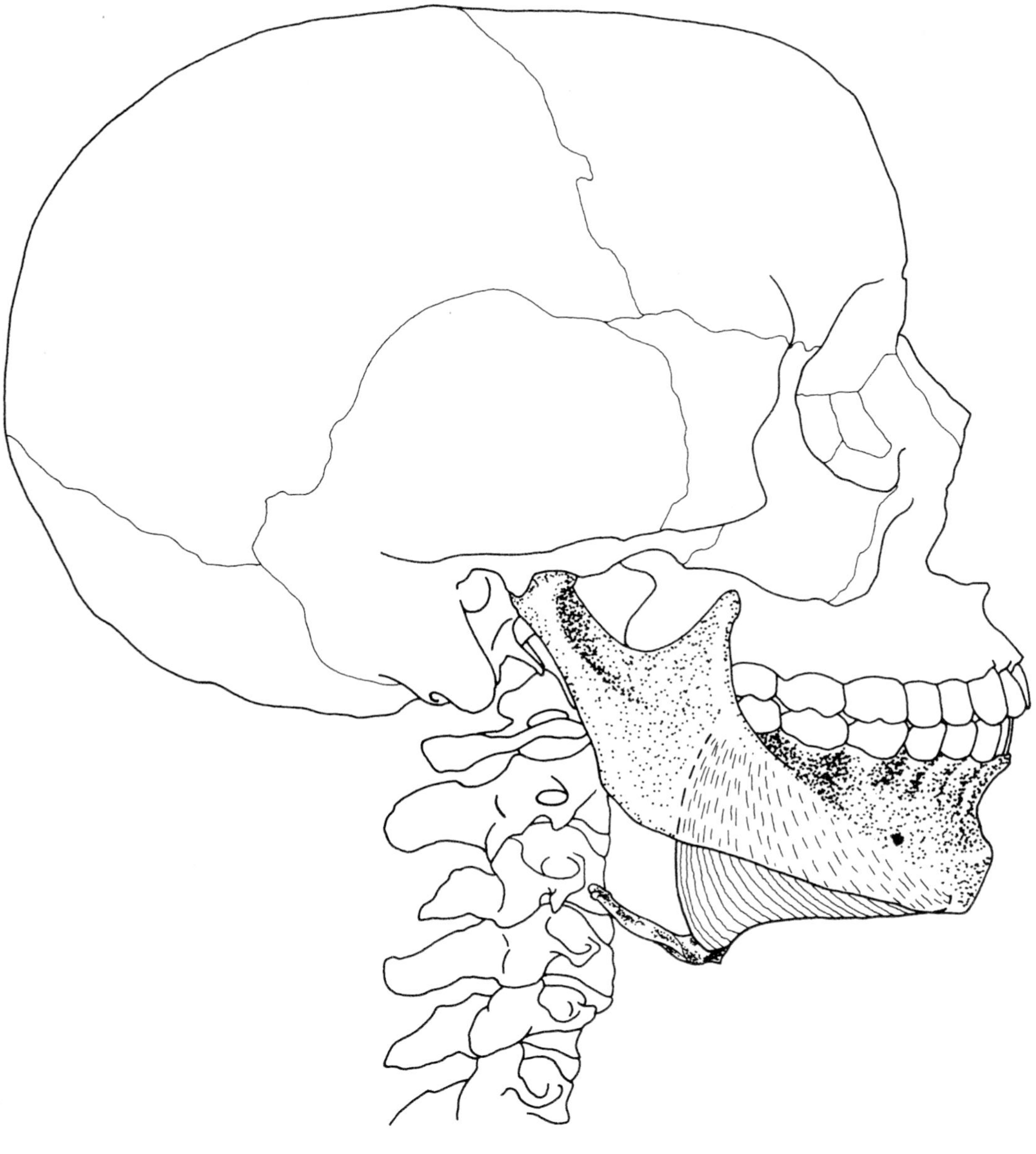

Lateral view

Origin	Inside surface of mandible from symphysis to molars (mylohyoid line)	**Action**	Elevates hyoid bone, raises floor of mouth and tongue
Insertion	Hyoid bone	**Nerve**	Mandibular division of trigeminal nerve

GENIOHYOIDEUS

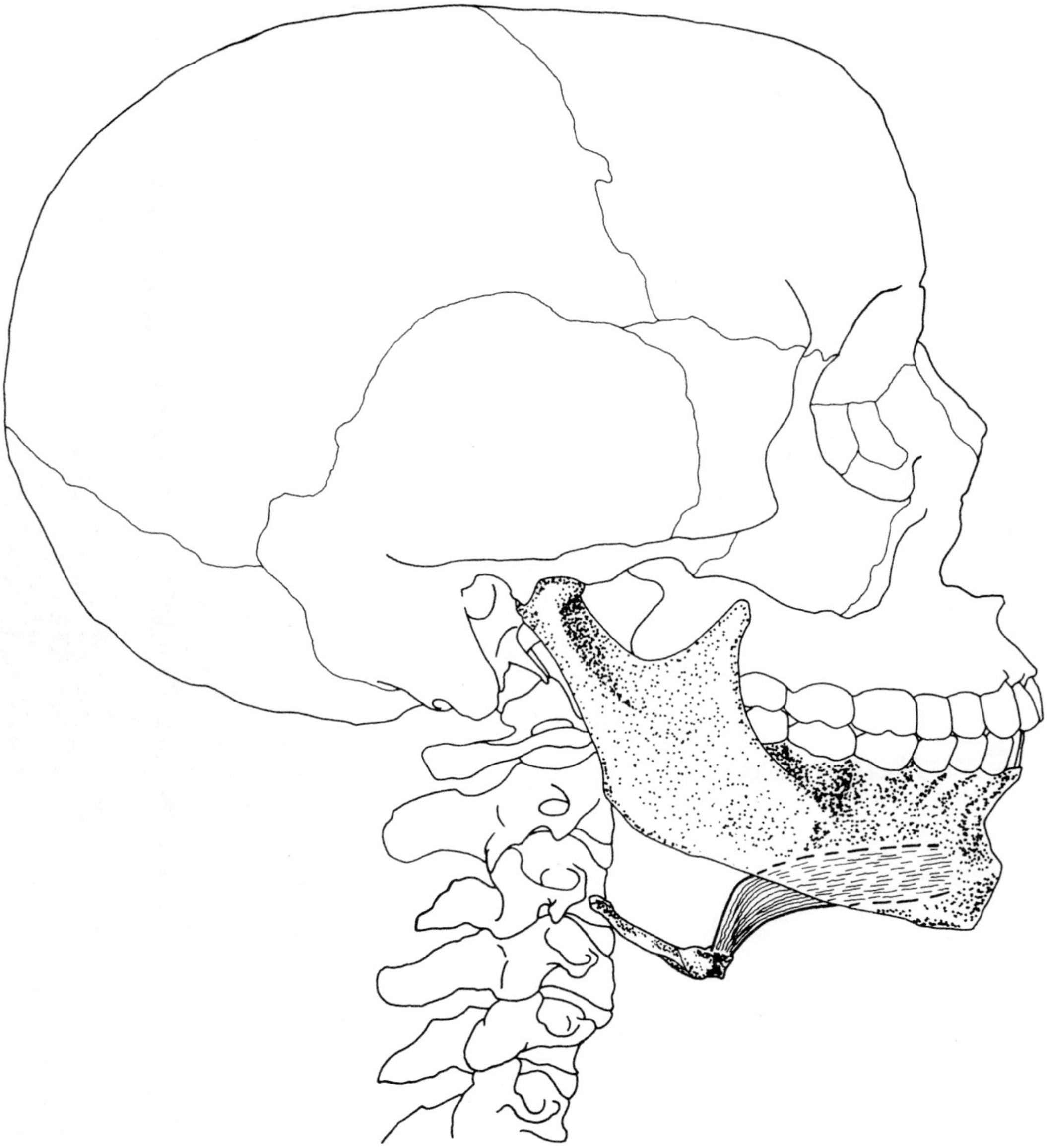

Lateral view

Origin Inferior mental spine on interior medial surface of mandible

Insertion Body of hyoid bone

Action Protrudes hyoid bone and tongue

Nerve Branch of C1 through hypoglossal nerve

STERNOHYOIDEUS

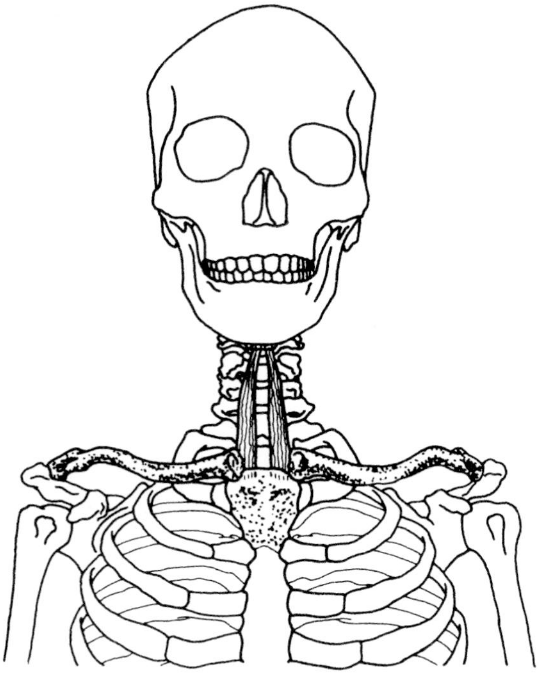

Frontal view

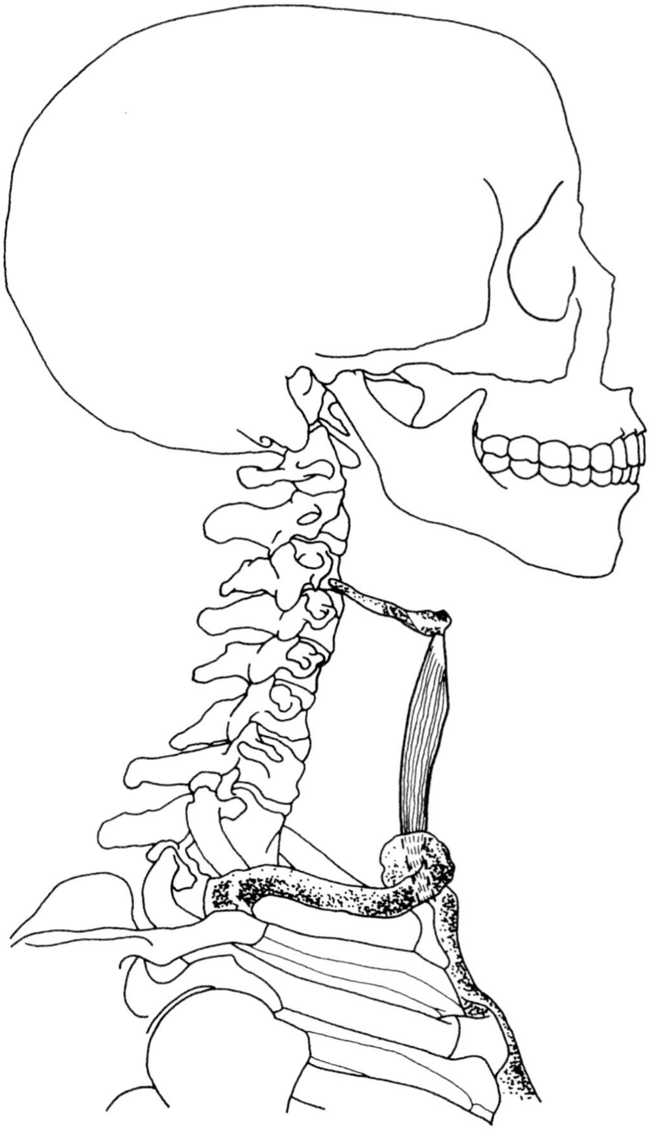

Lateral view

Origin	Medial end of clavicle, manubrium of sternum
Insertion	Body of hyoid bone
Action	Depresses hyoid bone
Nerve	Ansa cervicalis (C1–C3)

STERNOTHYROIDEUS

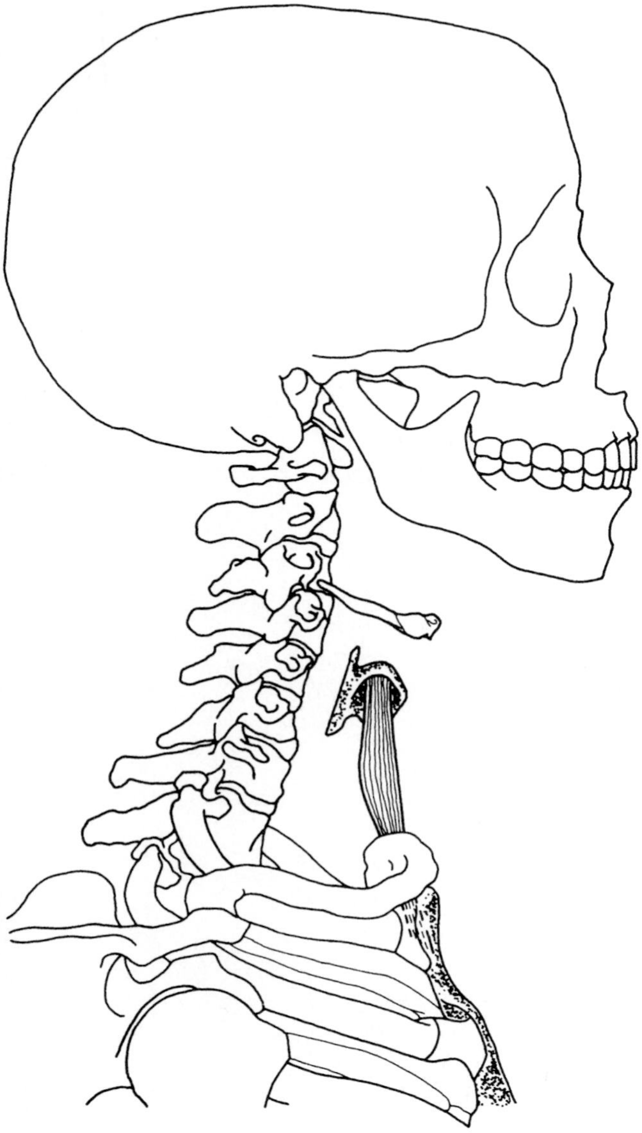

Lateral view

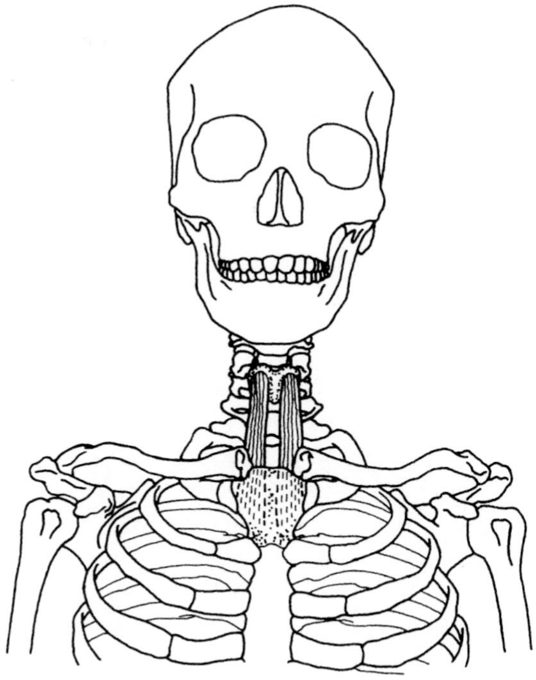

Frontal view

Origin	Dorsal surface of manubrium of sternum
Insertion	Lamina of thyroid cartilage
Action	Depresses thyroid cartilage
Nerve	Ansa cervicalis (C1–C3)

THYROHYOIDEUS

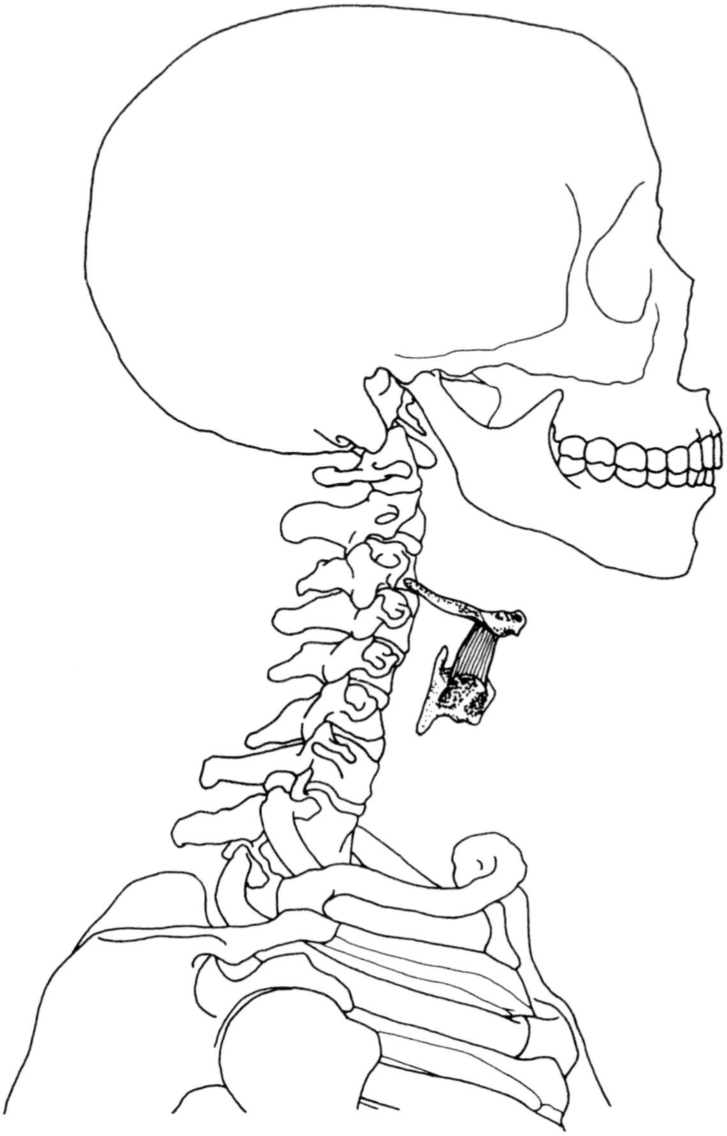

Lateral view

Origin	Lamina of thyroid cartilage	**Action**	Depresses hyoid or raises thyroid
Insertion	Greater cornu of hyoid bone	**Nerve**	C1 through hypoglossal nerve

OMOHYOIDEUS

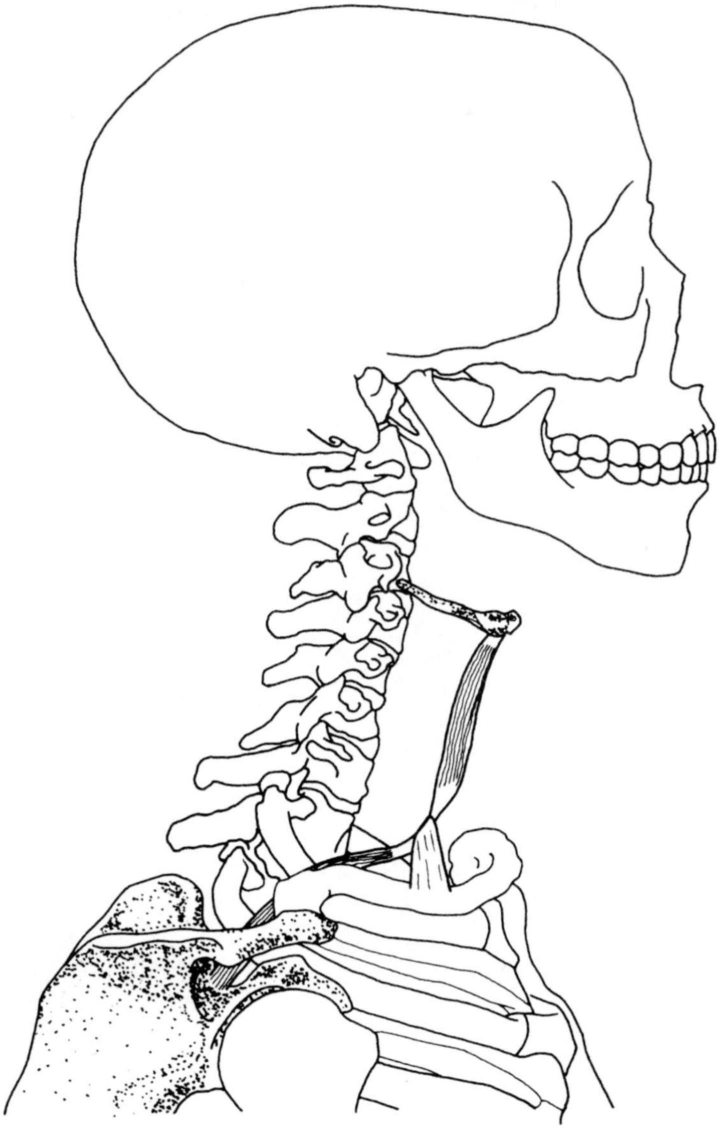

Lateral view

Origin Superior border of scapula

Insertion Inferior belly—bound to clavicle by central tendon
Superior belly—continues to body of hyoid bone

Action Depresses hyoid bone

Nerve Ansa cervicalis (C2, C3)

LONGUS COLLI

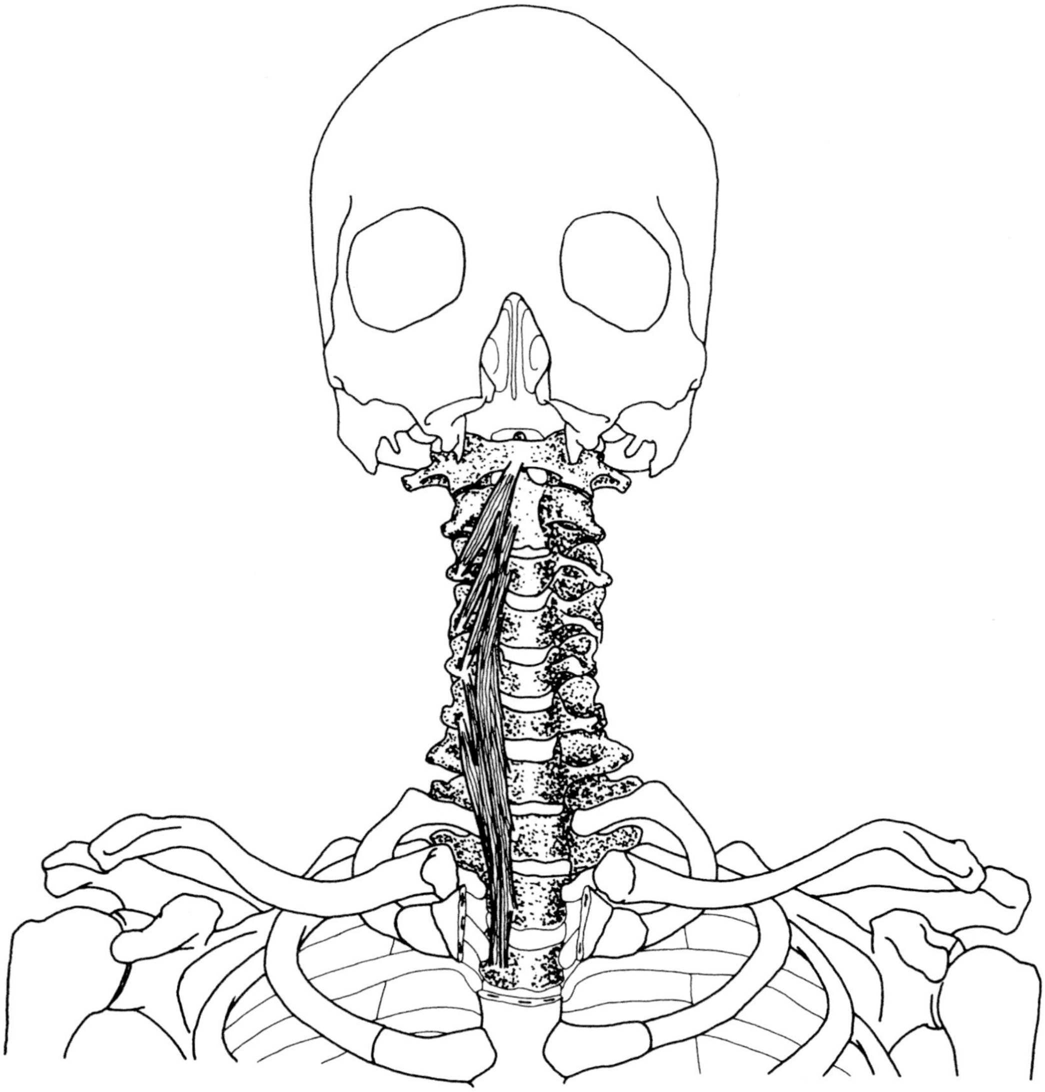

Frontal view

(Mandible and part of maxilla removed)

Superior oblique part

Origin	Transverse processes of third, fourth, and fifth cervical vertebrae
Insertion	Anterior arch of atlas

Inferior oblique part

Origin	Anterior surface of bodies of first two or three thoracic vertebrae
Insertion	Transverse processes of fifth and sixth cervical vertebrae

Vertical part

Origin	Anterior surfaces of bodies of upper three thoracic and lower three cervical vertebrae
Insertion	Anterior surfaces of the second, third, and fourth cervical vertebrae
Action	All three parts flex cervical vertebrae
Nerve	C2–C7

LONGUS CAPITIS

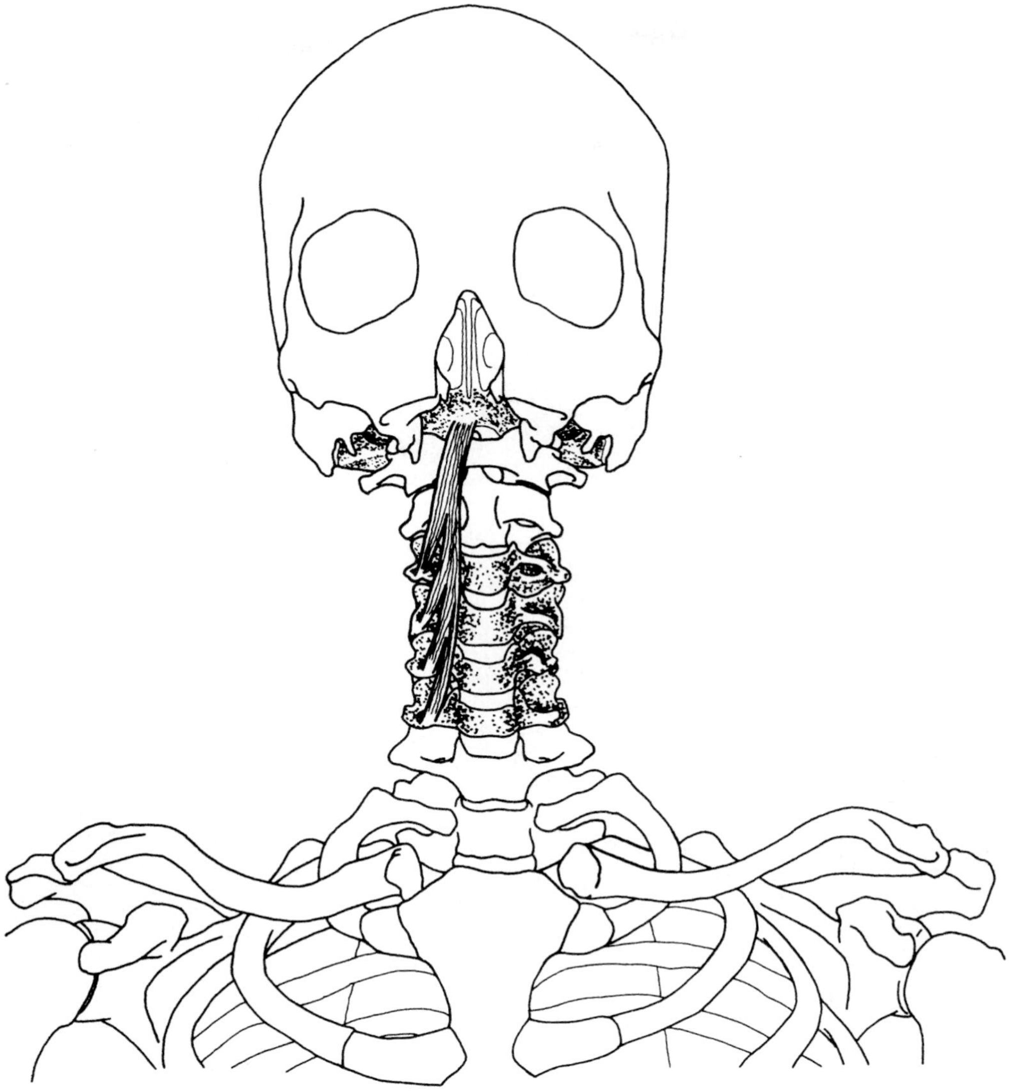

Frontal view

(Mandible and part of maxilla removed)

Origin Transverse processes of third through sixth cervical vertebrae

Insertion Occipital bone anterior to foramen magnum

Action Flexes head

Nerve C1–C3

RECTUS CAPITIS ANTERIOR

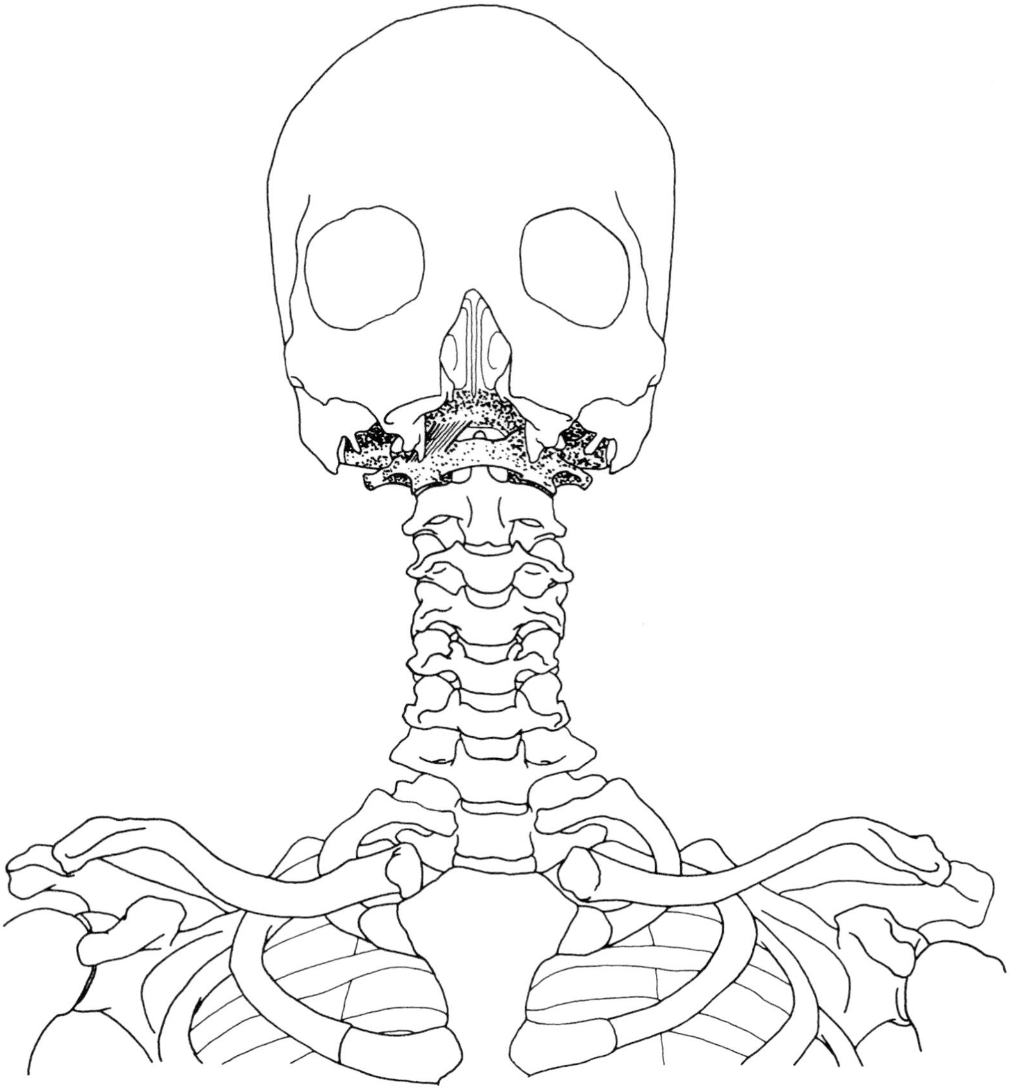

Frontal view

(Mandible and part of maxilla removed)

Origin Anterior base of transverse process of atlas

Insertion Occipital bone anterior to foramen magnum

Action Flexes head

Nerve C2, C3

RECTUS CAPITIS LATERALIS

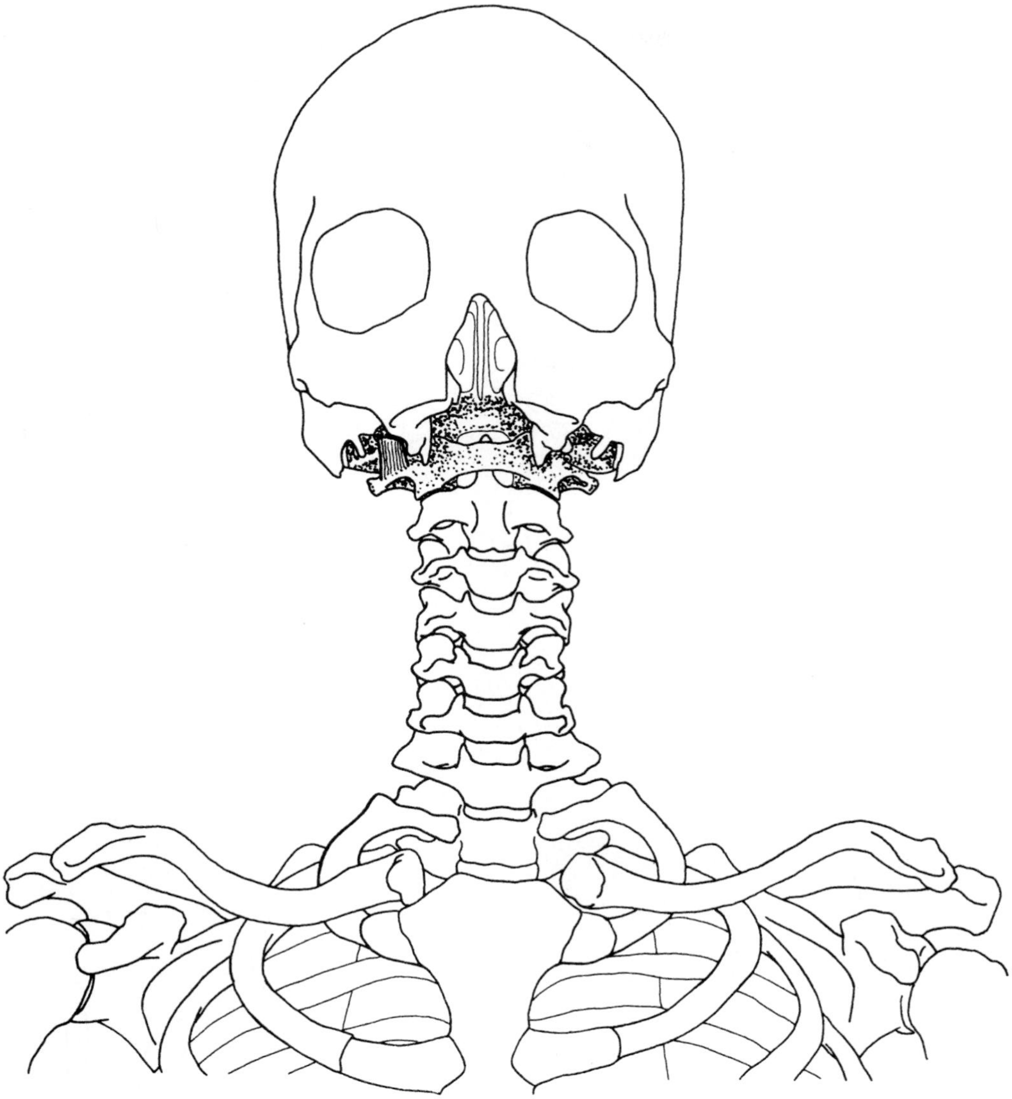

Frontal view

(Mandible and part of maxilla removed)

Origin Transverse process of atlas

Insertion Jugular process of occipital bone

Action Bends head laterally

Nerve C2, C3

SCALENUS ANTERIOR

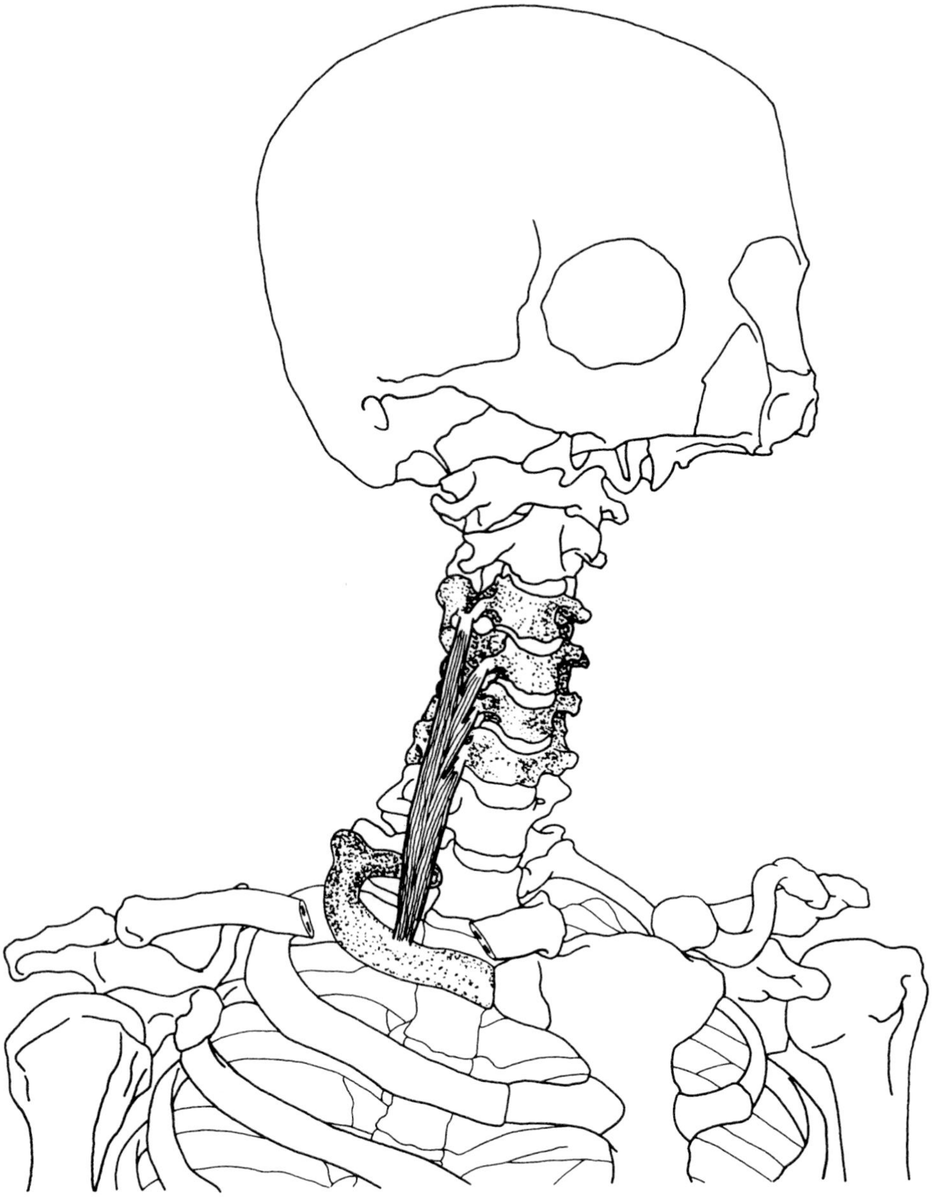

Three-quarter frontal view

(Mandible and part of maxilla removed)

Origin Transverse processes of third through sixth cervical vertebrae

Insertion Inner border of first rib (scalene tubercle)

Action Raises first rib (respiratory inspiration); acting together, they flex neck; acting on one side, they laterally flex, rotate neck

Nerve Ventral rami of cervical nerves

SCALENUS MEDIUS

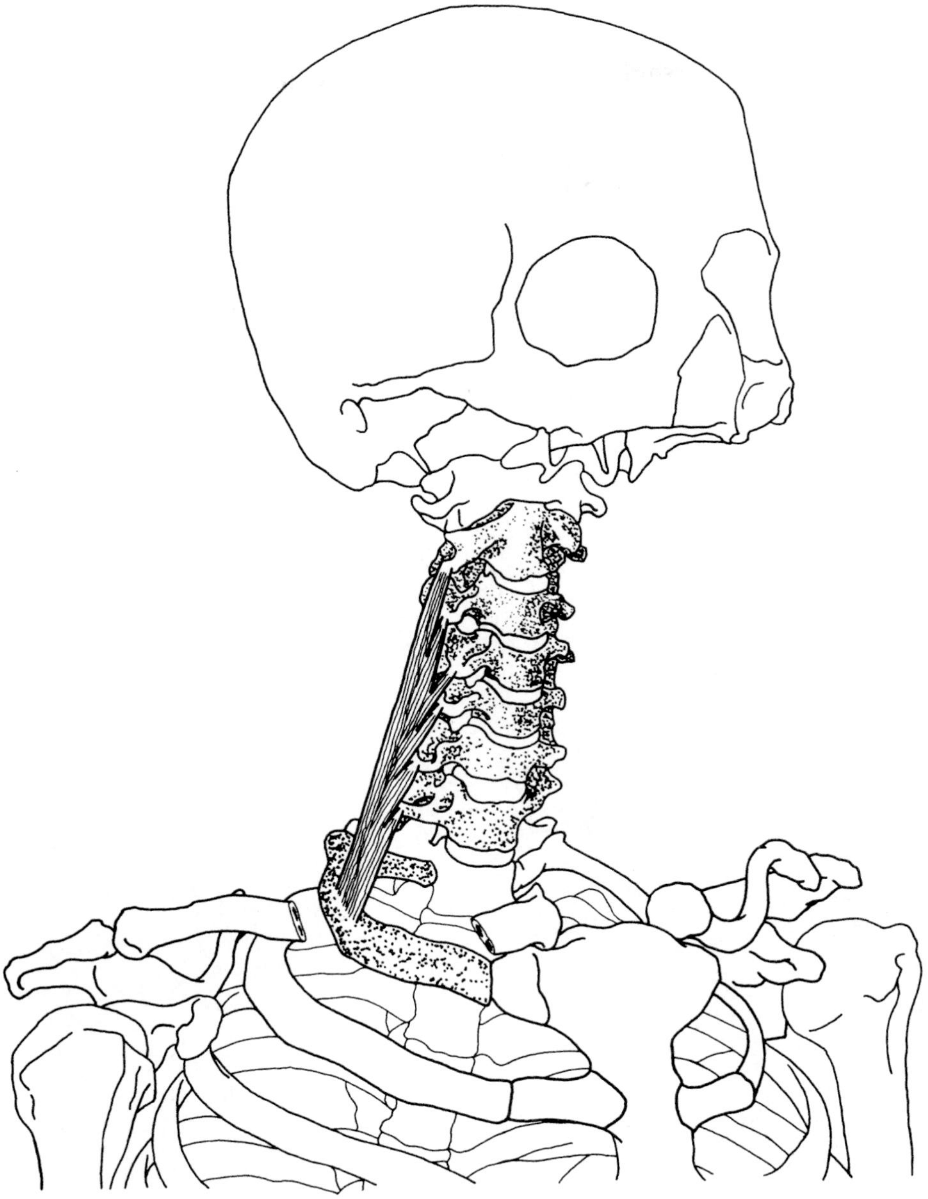

Three-quarter frontal view

(Mandible and part of maxilla removed)

Origin Transverse processes of lower six cervical vertebrae (C2–C7)

Insertion Upper surface of first rib

Action Raises first rib (respiratory inspiration); acting together, they flex neck; acting on one side, they laterally flex, rotate neck

Nerve Ventral rami of cervical nerves

SCALENUS POSTERIOR

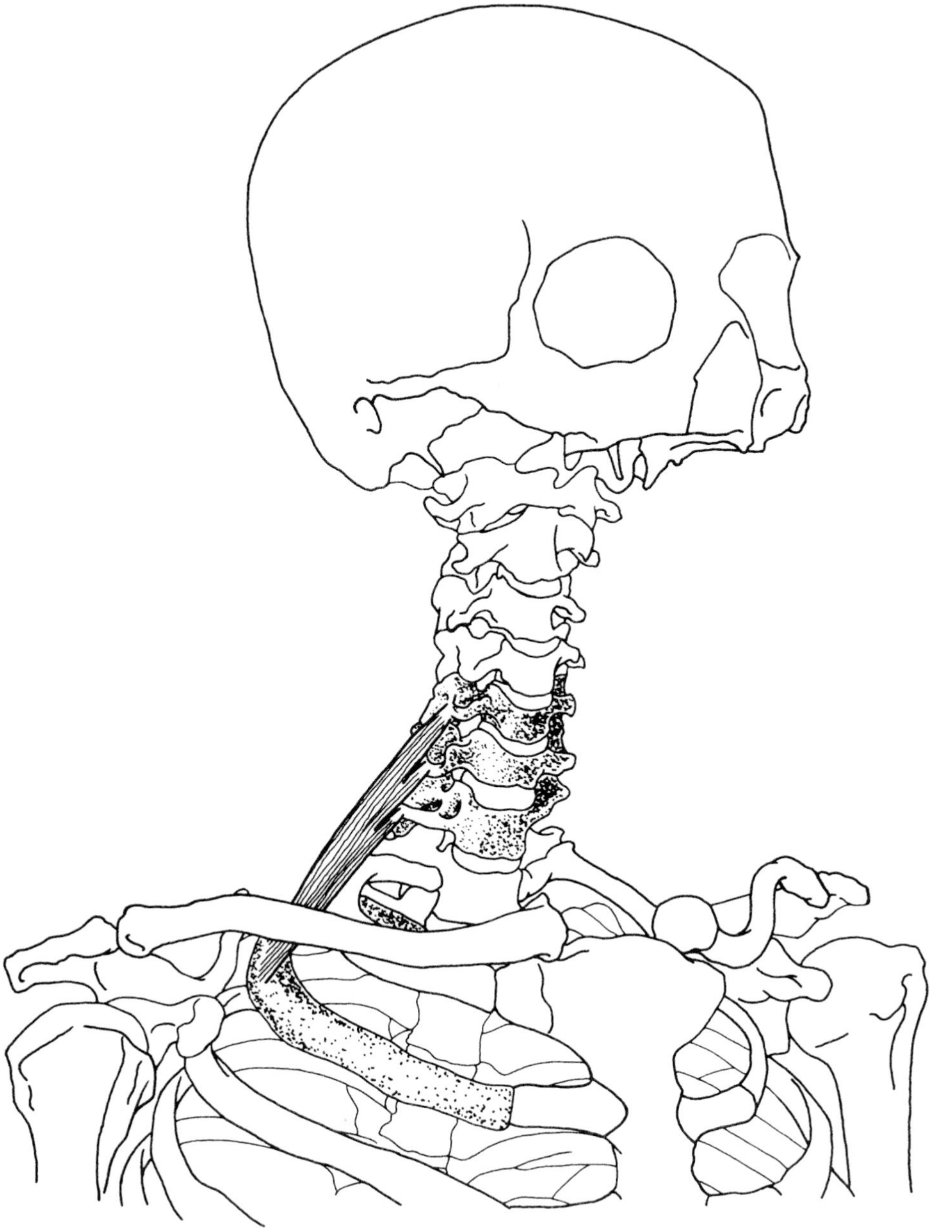

Three-quarter frontal view

(Mandible and part of maxilla removed)

Origin Transverse processes of lower two or three cervical vertebrae (C5–C7)

Insertion Outer surface of second rib

Action Raises second rib (respiratory inspiration); acting together, they flex neck; acting on one side, they laterally flex, rotate neck

Nerve Ventral rami of lower cervical nerves

RECTUS CAPITIS POSTERIOR MAJOR

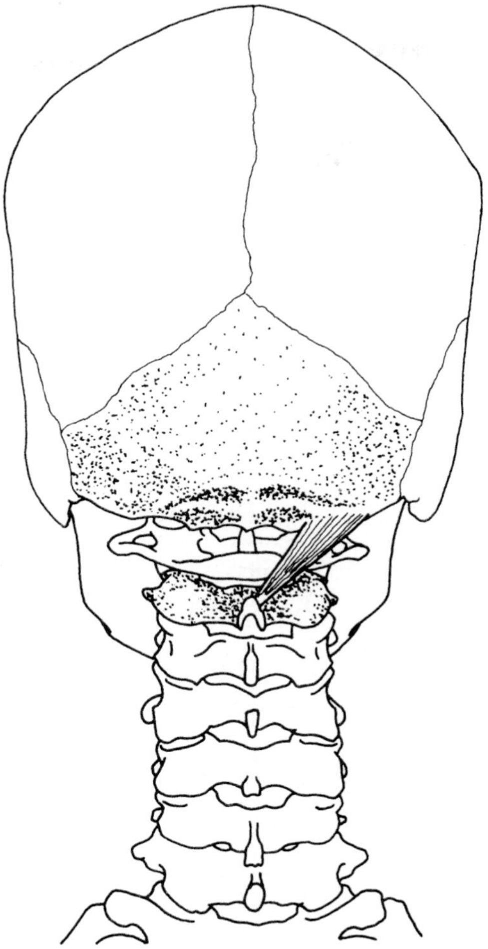

Posterior skull and cervical vertebrae

Origin Spinous process of axis

Insertion Lateral portion of inferior nuchal line of occipital bone

Action Extends and rotates head

Nerve Suboccipital nerve

RECTUS CAPITIS POSTERIOR MINOR

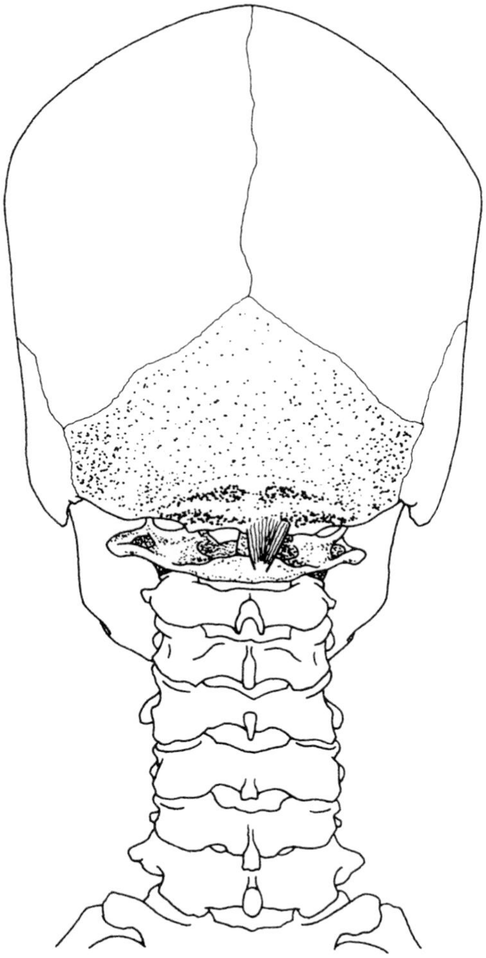

Posterior skull and cervical vertebrae

Origin Posterior arch of atlas

Insertion Medial portion of inferior nuchal line of occipital bone

Action Extends head

Nerve Suboccipital nerve

OBLIQUUS CAPITIS INFERIOR

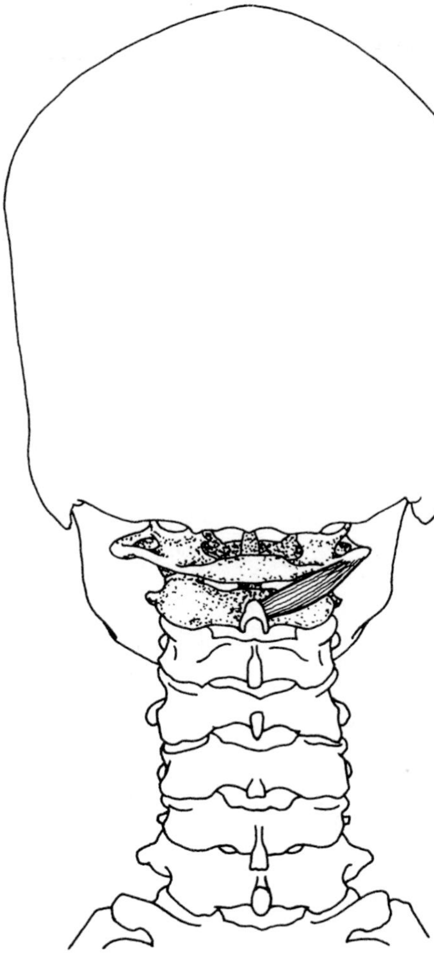

Posterior skull and cervical vertebrae

Origin	Spinous process of axis	**Action**	Rotates atlas
Insertion	Transverse process of atlas	**Nerve**	Suboccipital nerve

OBLIQUUS CAPITIS SUPERIOR

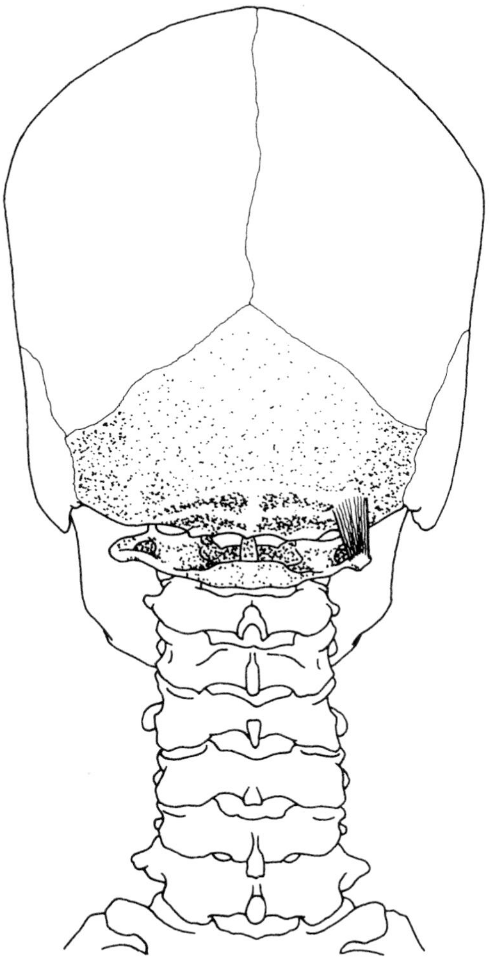

Posterior skull and cervical vertebrae

Origin Transverse process of atlas

Insertion Occipital bone between inferior and superior nuchal lines

Action Extends and bends head laterally

Nerve Suboccipital nerve

CHAPTER FIVE

MUSCLES OF THE TRUNK

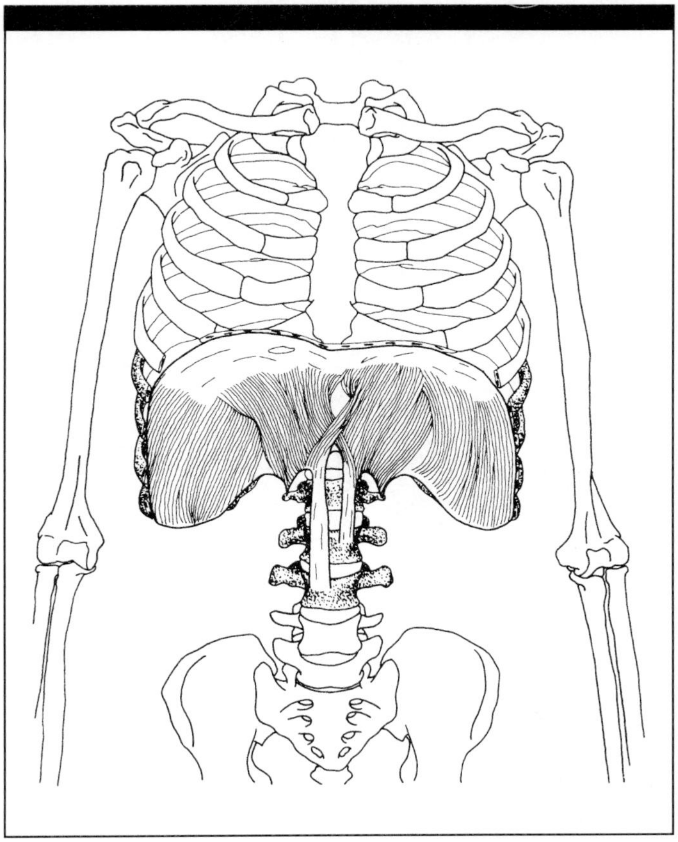

SPLENIUS CAPITIS

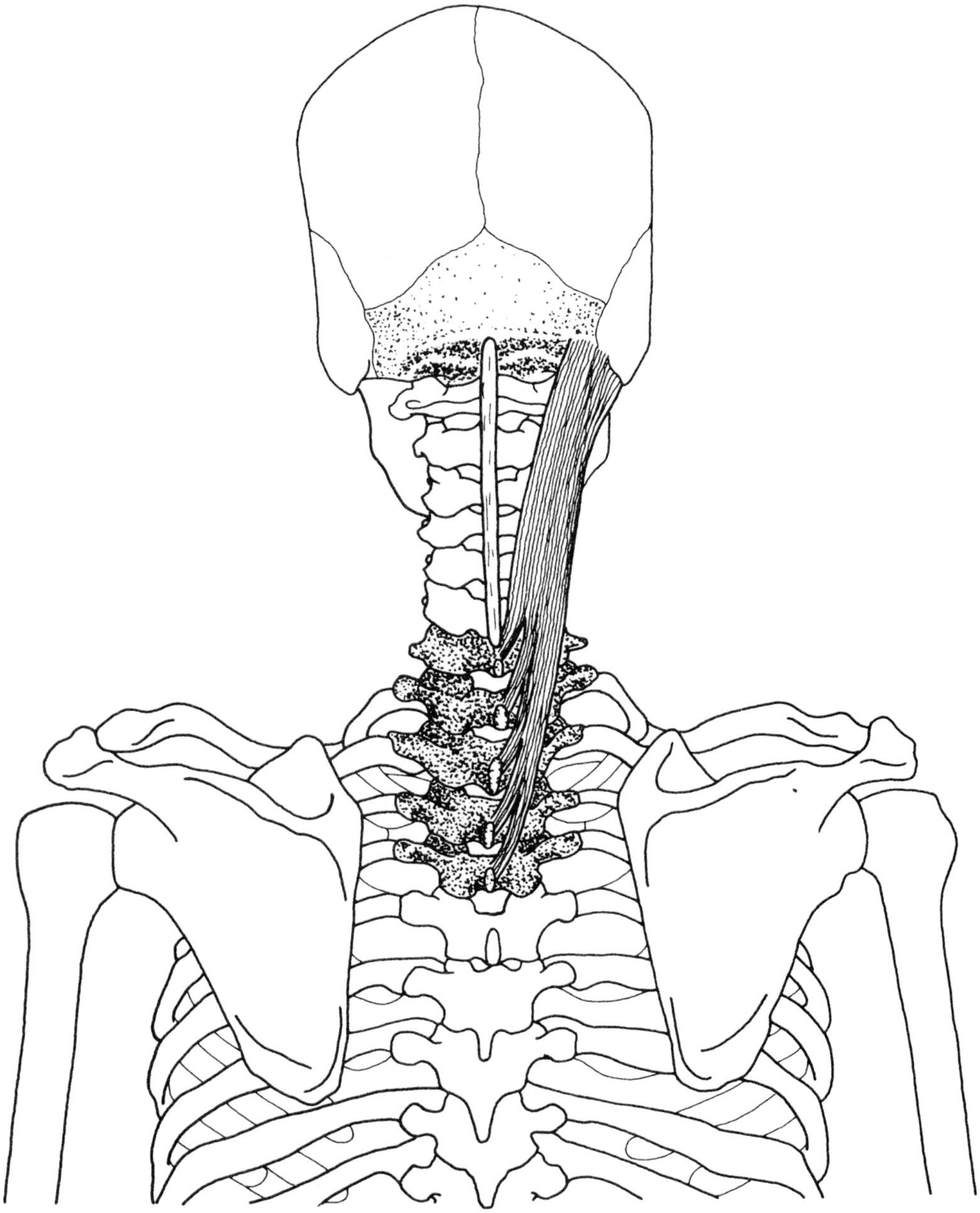

Posterior skull, neck, and back

Origin
Lower part of ligamentum nuchae, spinous processes of seventh cervical vertebra (C7) and upper three or four thoracic vertebrae (T1–T4)

Insertion
Mastoid process of temporal bone and lateral part of superior nuchal line

Action
Acting together, they extend, hyperextend head, neck; acting on one side, they laterally flex, rotate head, neck

Nerve
Lateral branches of dorsal primary divisions of middle and lower cervical nerves

SPLENIUS CERVICIS

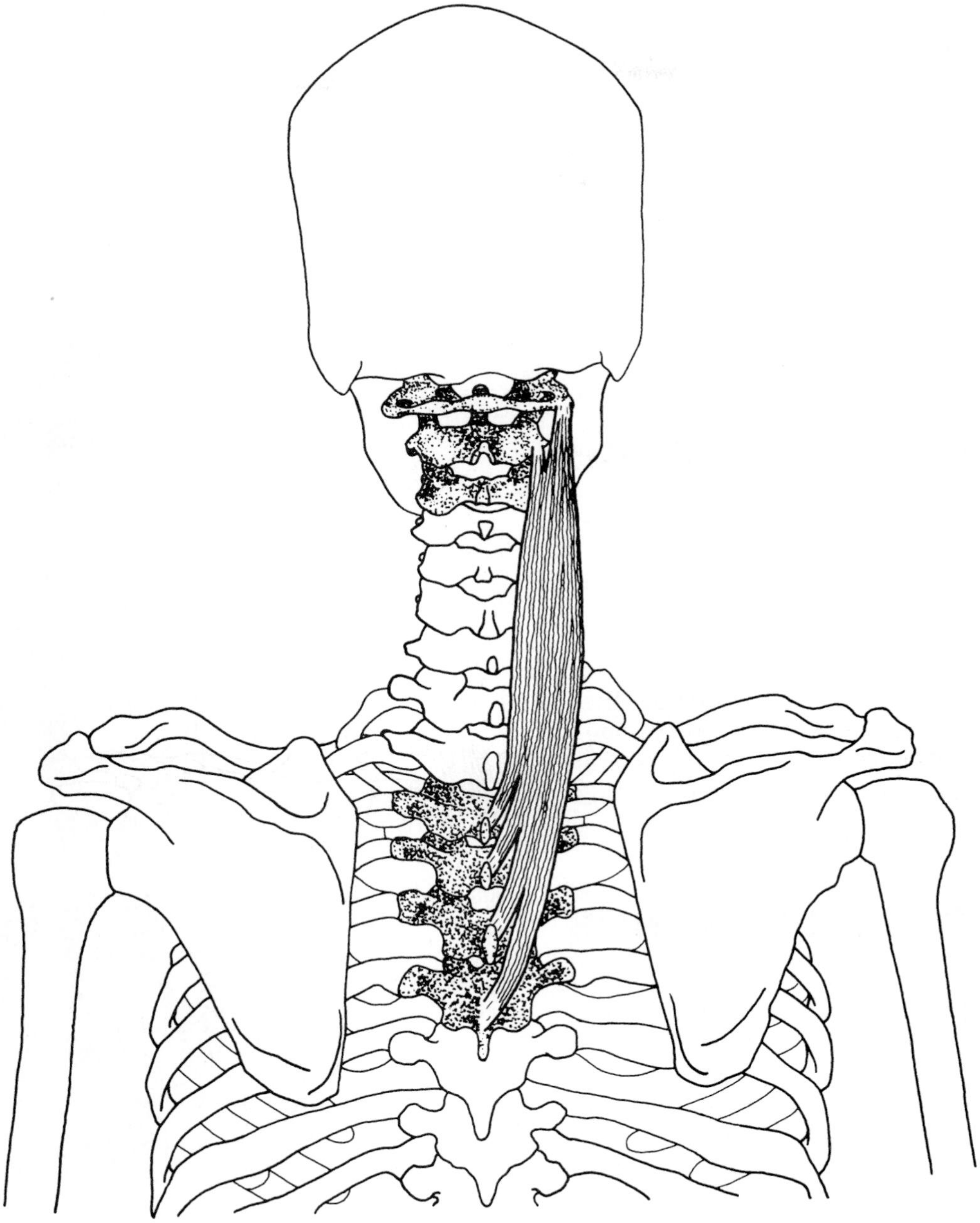

Posterior skull, neck, and back

Origin Spinous processes of third through sixth thoracic vertebrae (T3–T6)

Insertion Transverse processes of upper two or three cervical vertebrae (C1–C3)

Action Acting together, they extend, hyperextend head, neck; acting on one side, they laterally flex, rotate head, neck

Nerve Lateral branches of dorsal primary divisions of middle and lower cervical nerves

ERECTOR SPINAE*

ILIOCOSTALIS LUMBORUM

Origin Medial and lateral sacral crests and medial part of iliac crests
Insertion Angles of lower six ribs
Action Extension, lateral flexion of vertebral column, rotates ribs for forceful inspiration
Nerve Dorsal primary divisions of spinal nerves

ILIOCOSTALIS THORACIS

Origin Angles of lower six ribs medial to iliocostalis lumborum
Insertion Angles of upper six ribs and transverse process of seventh cervical vertebra
Action Extension, lateral flexion of vertebral column, rotates ribs for forceful inspiration
Nerve Dorsal primary divisions of spinal nerves

ILIOCOSTALIS CERVICIS

Origin Angles of third through sixth ribs
Insertion Transverse processes of fourth, fifth, and sixth cervical vertebrae
Action Extension, lateral flexion of vertebral column
Nerve Dorsal primary divisions of spinal nerves

*The erector spinae (sacrospinalis) is a complex of three sets of muscles: iliocostalis, longissimus, and spinalis. The origin of this group is the medial and lateral sacral crests, the medial part of iliac crests, and the spinous processes and supraspinal ligament of lumbar and eleventh and twelfth thoracic vertebrae.

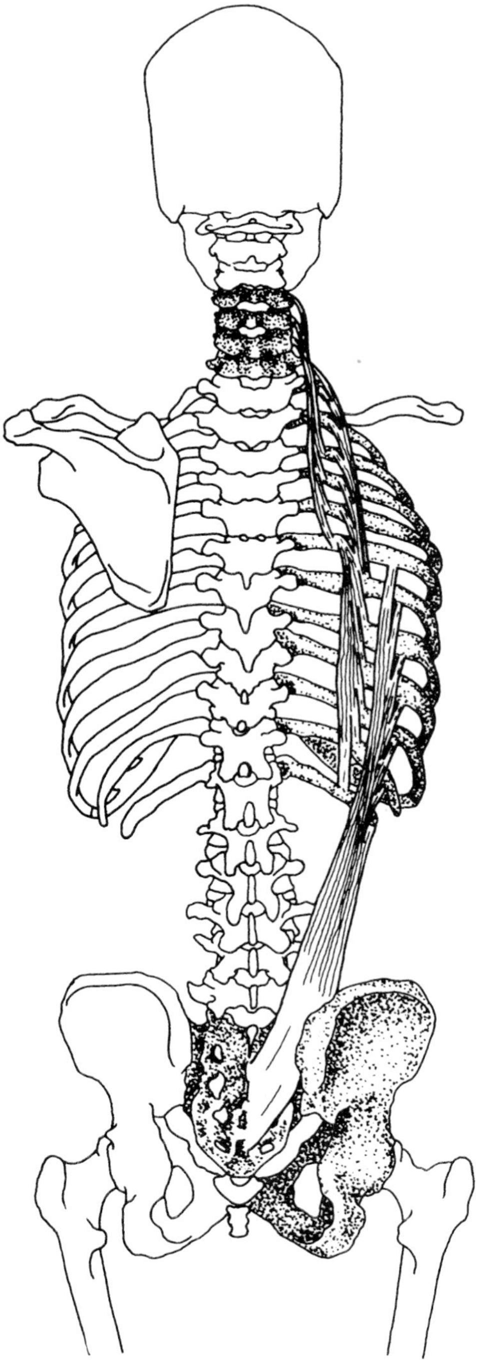

Trunk—dorsal view

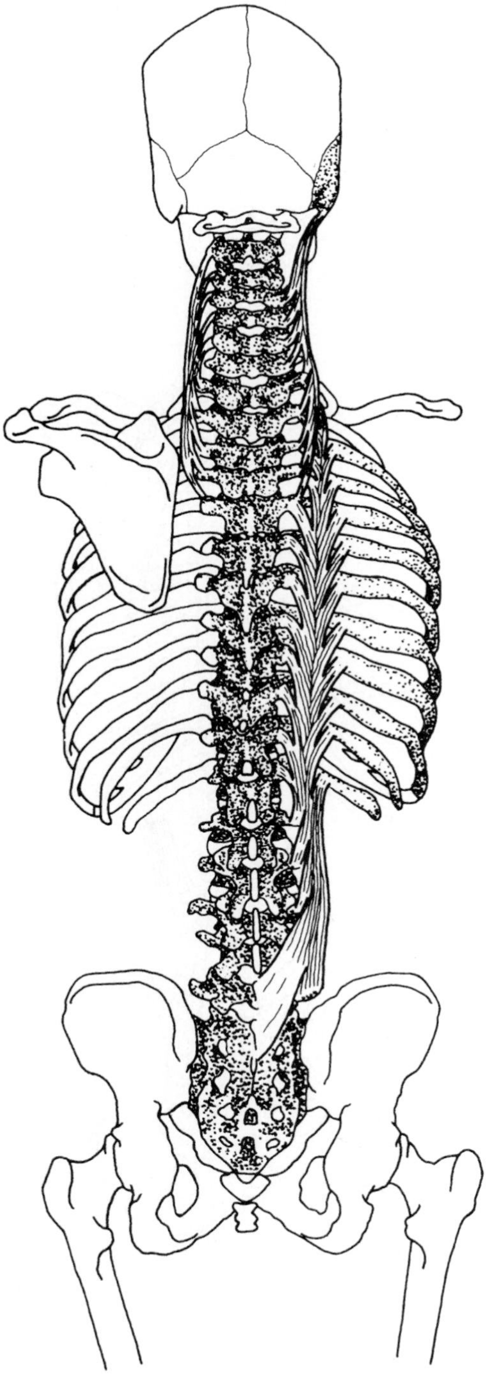

Trunk—dorsal view

ERECTOR SPINAE

LONGISSIMUS THORACIS

Origin	Medial and lateral sacral crests, spinous processes and supraspinal ligament of lumbar and eleventh and twelfth thoracic vertebrae, and medial part of iliac crests
Insertion	Transverse processes of all thoracic vertebrae, between tubercles and angles of lower nine or ten ribs
Action	Extension, lateral flexion of vertebral column, rotates ribs for forceful inspiration
Nerve	Dorsal primary divisions of spinal nerves

LONGISSIMUS CERVICIS

Origin	Transverse processes of upper four or five thoracic vertebrae (T1–T5)
Insertion	Transverse processes of second through sixth cervical vertebrae
Action	Extension, lateral flexion of vertebral column
Nerve	Dorsal primary divisions of spinal nerves

LONGISSIMUS CAPITIS

Origin	Transverse processes of upper five thoracic vertebrae (T1–T5), articular processes of lower three cervical vertebrae (C5–C7)
Insertion	Posterior part of mastoid process of temporal bone
Action	Extends and rotates head
Nerve	Dorsal primary divisions of middle and lower cervical nerves

ERECTOR SPINAE

SPINALIS THORACIS

Origin Spinous processes of lower two thoracic (T11, T12) and upper two lumbar (L1, L2) vertebrae

Insertion Spinous processes of upper thoracic vertebrae (T1–T8)

Action Extends vertebral column

Nerve Dorsal primary divisions of spinal nerves

SPINALIS CERVICIS

Origin Ligamentum nuchae, spinous process of seventh cervical vertebra

Insertion Spinous process of axis

Action Extends vertebral column

Nerve Dorsal primary divisions of spinal nerves

SPINALIS CAPITIS

(Medial part of semispinalis capitis)

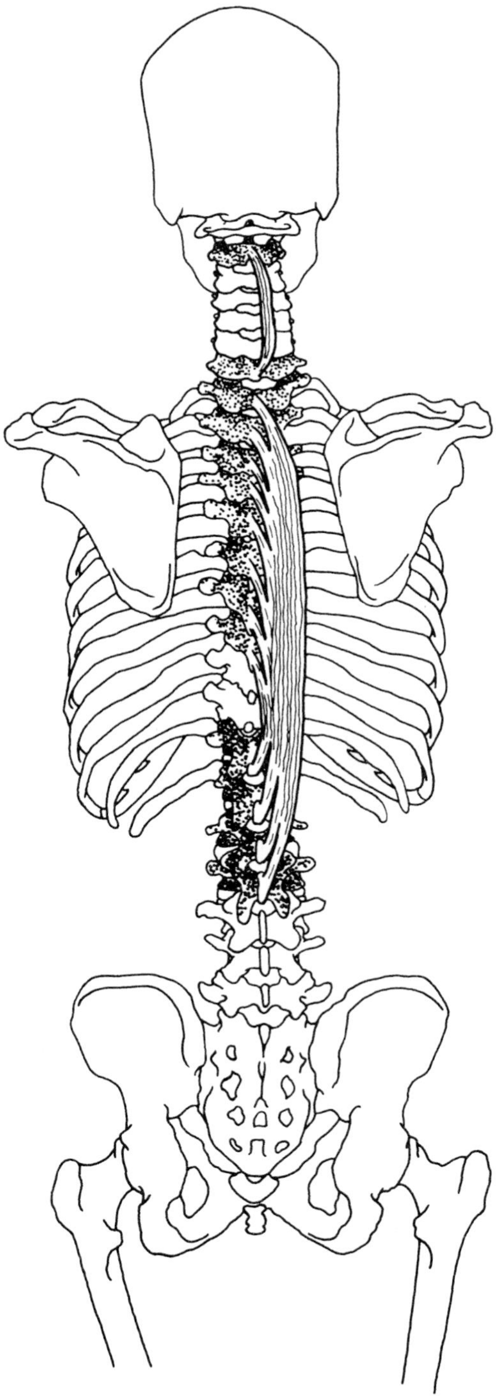

Trunk—dorsal view

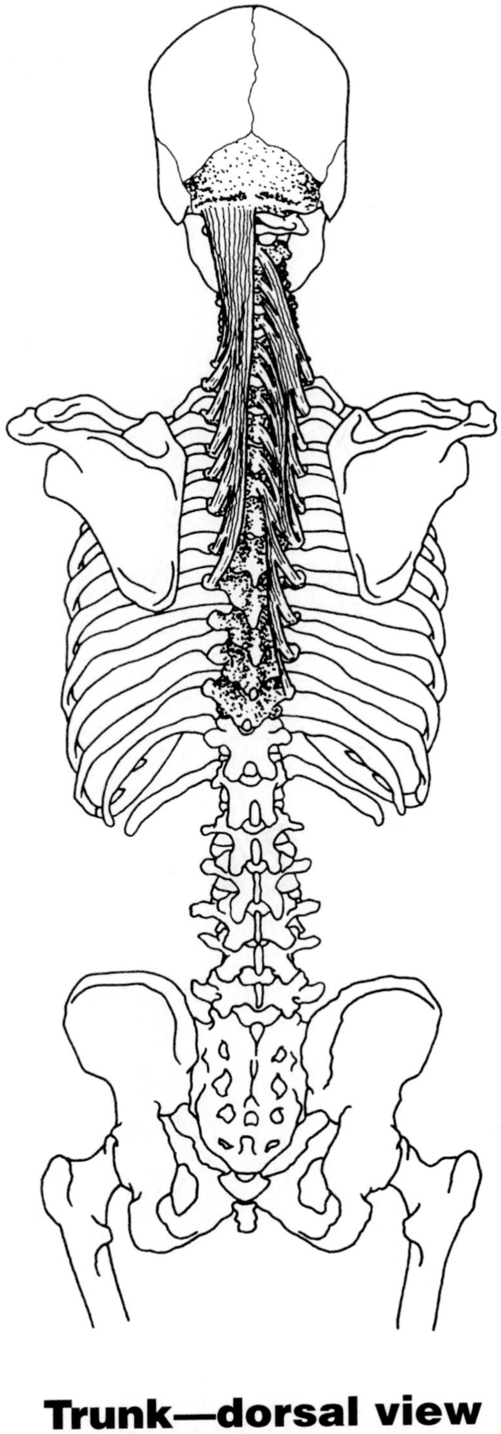

Trunk—dorsal view

TRANSVERSOSPINALIS*

SEMISPINALIS THORACIS

Origin	Transverse processes of the sixth through tenth thoracic vertebrae (T6–T10)
Insertion	Spinous processes of the lower two cervical (C6, C7) and upper four thoracic (T1–T4) vertebrae
Action	Extends and rotates vertebral column
Nerve	Dorsal primary divisions of spinal nerves

SEMISPINALIS CERVICIS

Origin	Transverse processes of upper five or six thoracic vertebrae (T1–T6)
Insertion	Spinous processes of second to fifth cervical vertebrae (C2–C5)
Action	Extends and rotates vertebral column
Nerve	Dorsal primary divisions of spinal nerves

SEMISPINALIS CAPITIS

(Medial part is spinalis capitis)

Origin	Transverse processes of lower four cervical (C4–C7) and upper six or seven thoracic (T1–T7) vertebrae
Insertion	Between superior and inferior nuchal lines of occipital bone
Action	Extends and rotates head
Nerve	Dorsal primary divisions of spinal nerves

*The transversospinalis is composed of groups of small muscles generally extending upward from transverse processes to spinous processes of higher vertebrae. They are deep to erector spinae. They include semispinalis, multifidi, and rotatores.

MULTIFIDIS*

Origin	Sacral region—along sacral foramina up to posterior superior iliac spine Lumbar region—mamillary processes† of vertebrae Thoracic region—transverse processes Cervical region—articular processes of lower four vertebrae (C4–C7)
Insertion	Spinous process two to four vertebrae superior to origin
Action	Extend and rotate vertebral column
Nerve	Dorsal primary division of spinal nerves

*Part of transversospinalis.

†Posterior border of superior articular process.

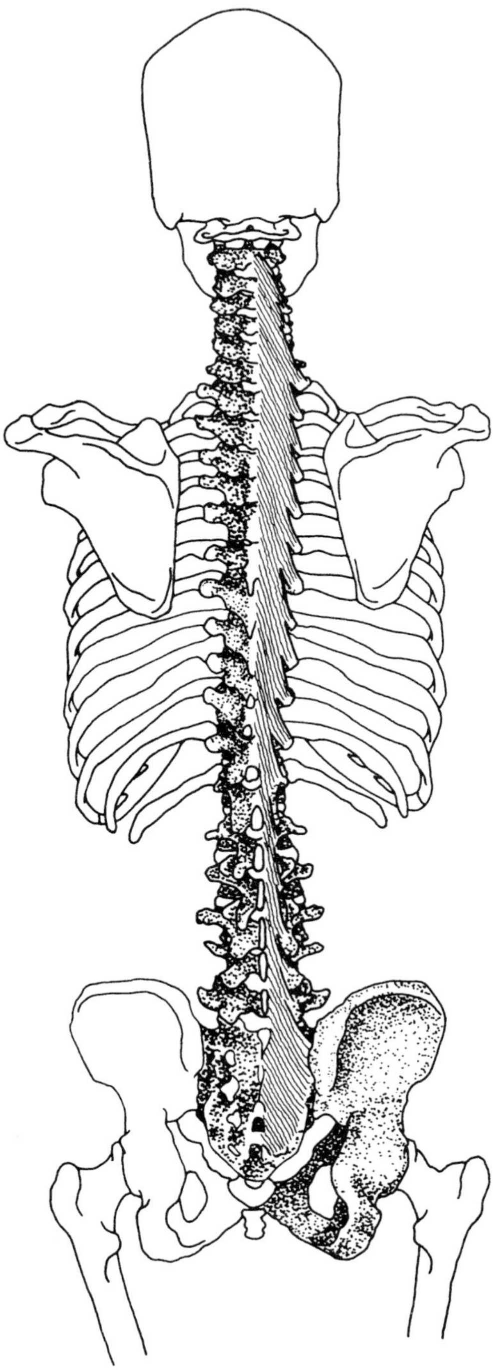

Trunk—dorsal view

ROTATORES*

Origin	Transverse process of each vertebra
Insertion	Base of spinous process of next vertebra above
Action	Extend and rotate vertebral column
Nerve	Dorsal primary division of spinal nerves

*Part of transversospinalis.

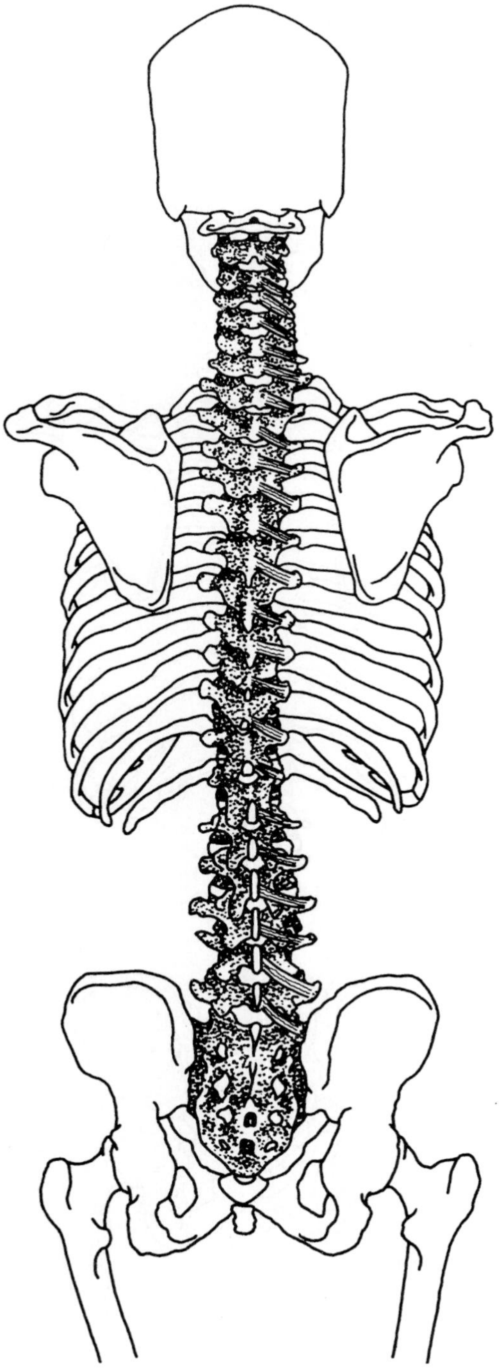

Trunk—dorsal view

INTERSPINALES

(Paired on either side of interspinal ligament)

Origin Cervical region—spinous processes of third to seventh cervical vertebrae (C3–C7)

Thoracic region—spinous processes of second to twelfth thoracic vertebrae (T2–T12)

Lumbar region—spinous processes of second to fifth lumbar vertebrae (L2–L5)

Insertion Spinous process of next vertebra superior to origin

Action Extend vertebral column

Nerve Dorsal primary division of spinal nerves

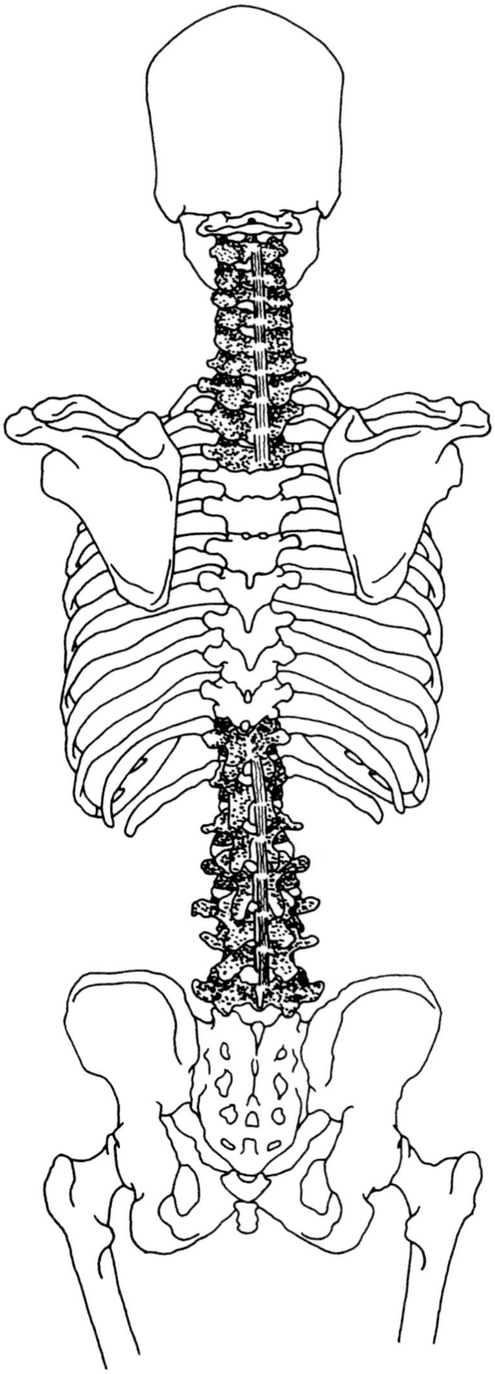

Trunk—dorsal view

INTERTRANSVERSARII

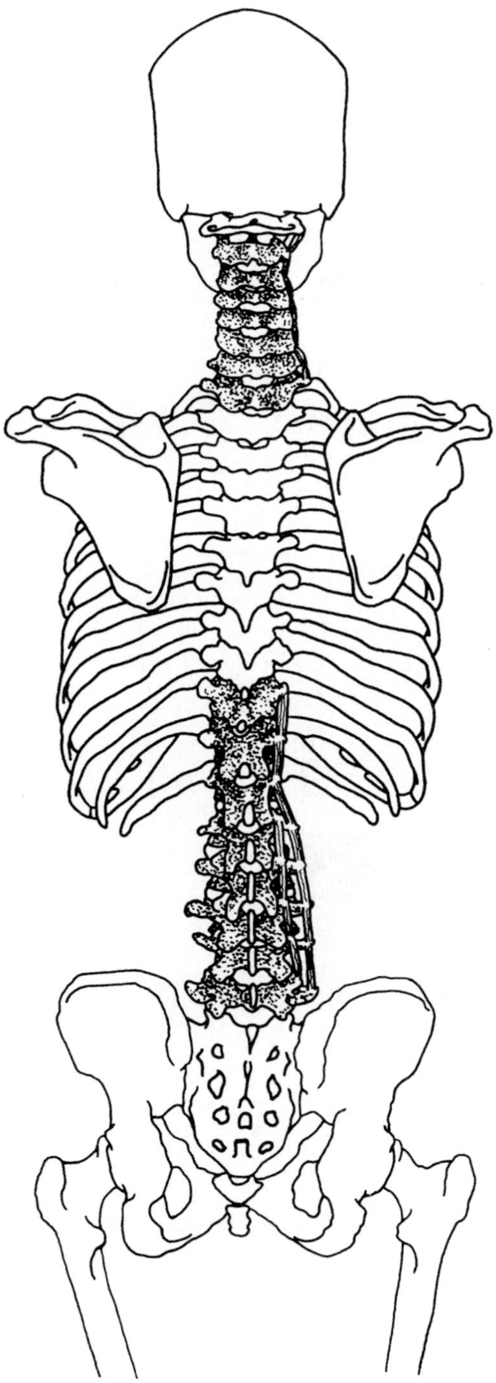

Trunk—dorsal view

Cervical region

INTERTRANSVERSARII ANTERIORES

Origin Anterior tubercle of transverse processes of vertebrae from first thoracic to axis

Insertion Anterior tubercle of next superior vertebra

Action Lateral flexion of vertebral column

Nerve Ventral primary division of spinal nerves

INTERTRANSVERSARII POSTERIORES

Origin Posterior tubercle of transverse processes of vertebrae from first thoracic to axis

Insertion Posterior tubercle of next superior vertebra

Thoracic region

Origin Transverse processes of first lumbar to eleventh thoracic vertebrae

Insertion Transverse processes of next superior vertebra

Lumbar region

INTERTRANSVERSARII LATERALES

Origin Transverse processes of lumbar vertebrae

Insertion Transverse process of next superior vertebra

Action Lateral flexion of vertebral column

Nerve Ventral primary division of spinal nerves

INTERTRANSVERSARII MEDIALES

Origin Mamillary process† of each lumbar vertebra

Insertion Accessory process of the next superior lumbar vertebra

Action Lateral flexion of vertebral column

Nerve Dorsal primary division of spinal nerves

†Posterior border of superior articular process.

INTERCOSTALES EXTERNI

(External Intercostal)

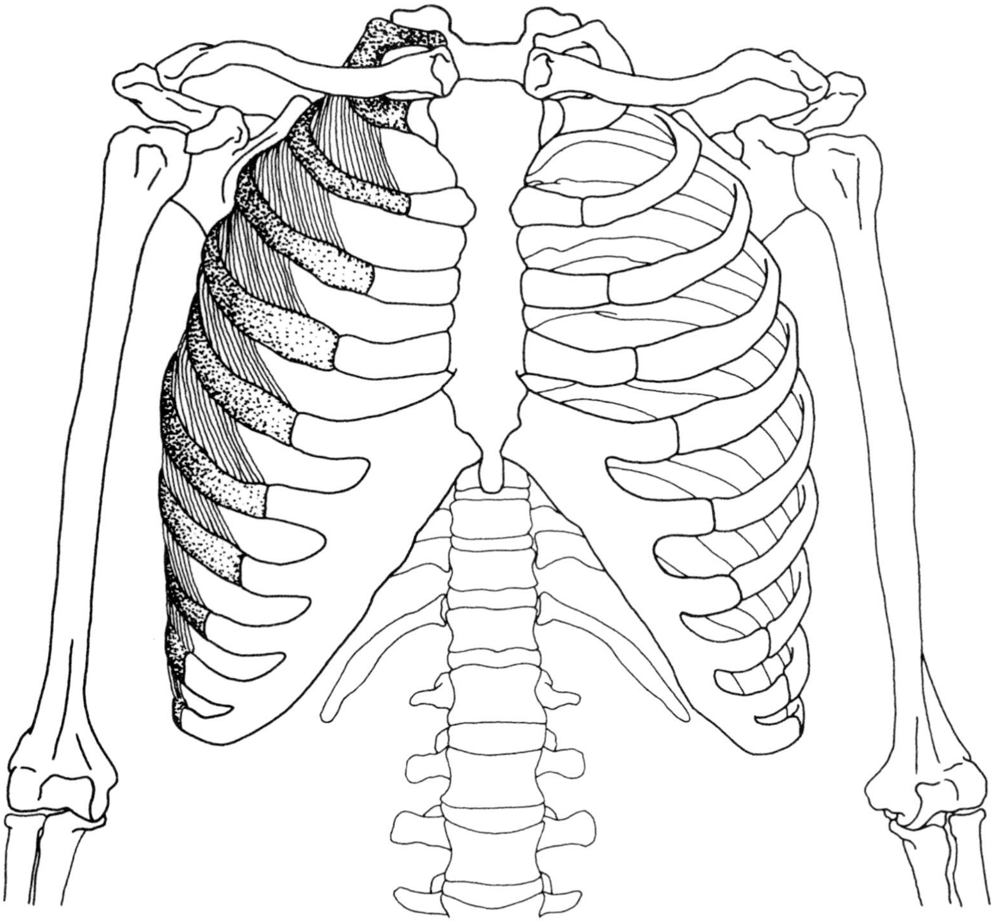

Trunk—anterior view

Origin Lower margin of upper eleven ribs

Insertion Superior border of rib below (each muscle fiber runs obliquely and inserts toward the costal cartilage)

Action Draw ventral part of ribs upward, increasing the volume of the thoracic cavity for inspiration

Nerve Intercostal nerves

INTERCOSTALES INTERNI

(Internal Intercostal)

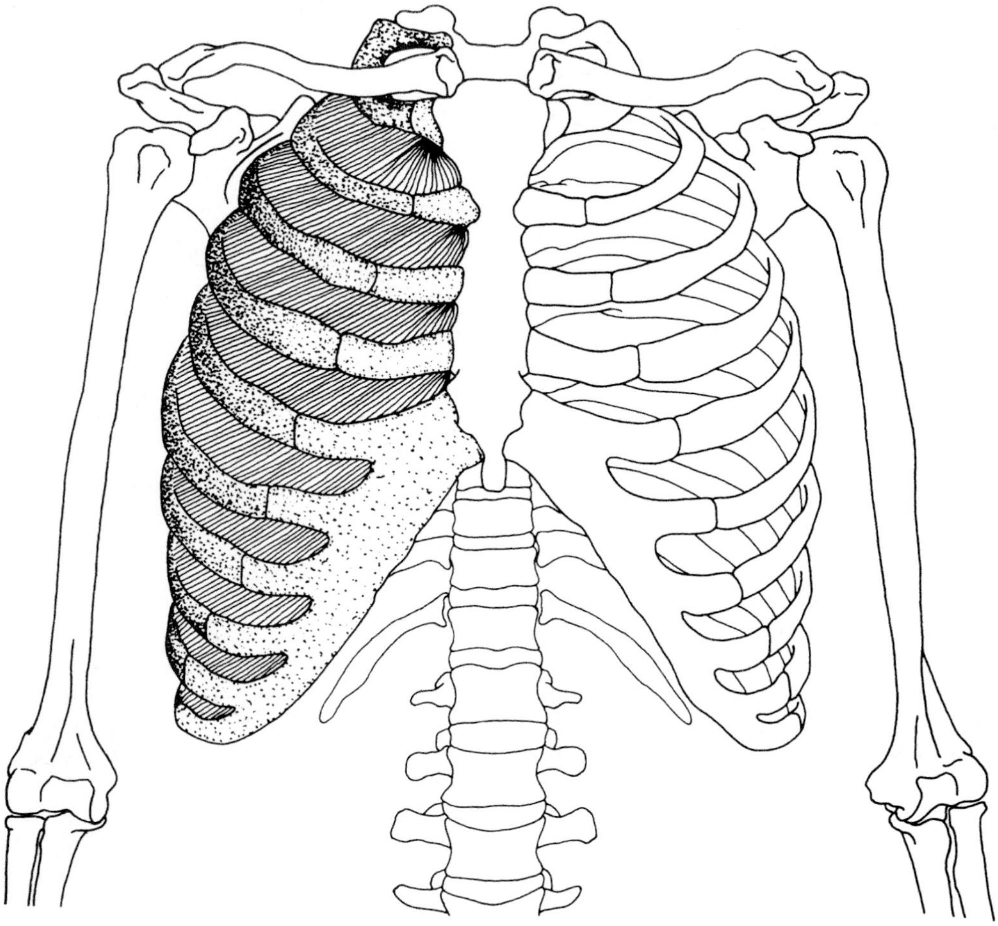

Trunk—anterior view

Origin From the cartilages to the angles of the upper eleven ribs

Insertion Superior border of the rib below (each muscle fiber runs obliquely and inserts away from the costal cartilage)

Action Draw ventral part of ribs downward, decreasing the volume of the thoracic cavity for expiration

Nerve Intercostal nerves

SUBCOSTALES

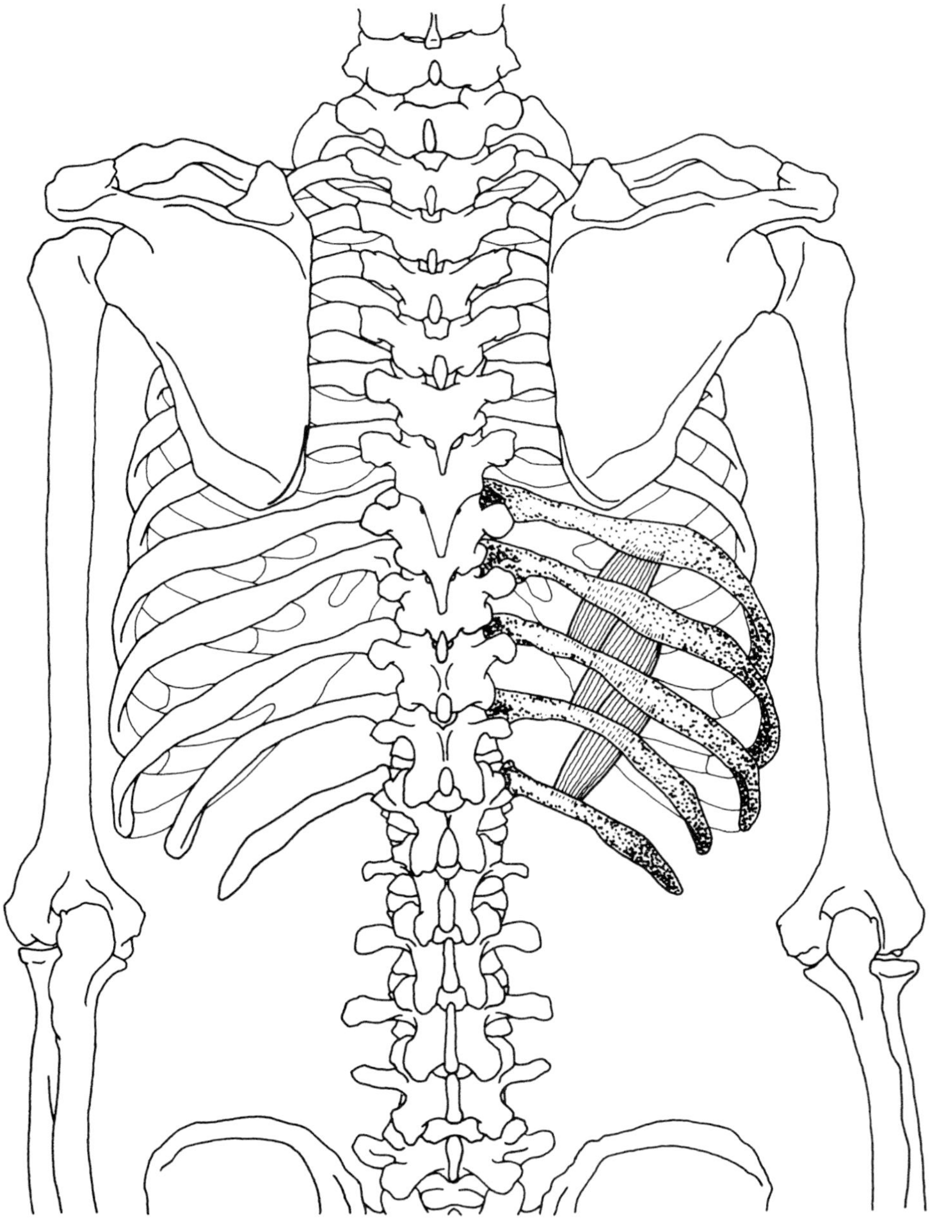

Trunk—dorsal view

Origin Inner surface of each rib near its angle

Insertion Medially on the inner surface of second or third rib below

Action Draw ventral part of ribs downward, decreasing the volume of the thoracic cavity for forceful expiration

Nerve Intercostal nerves

Note: These muscles are deep to the internal intercostals. They continue distally between single ribs, where they are known as innermost intercostal muscles.

TRANSVERSUS THORACIS

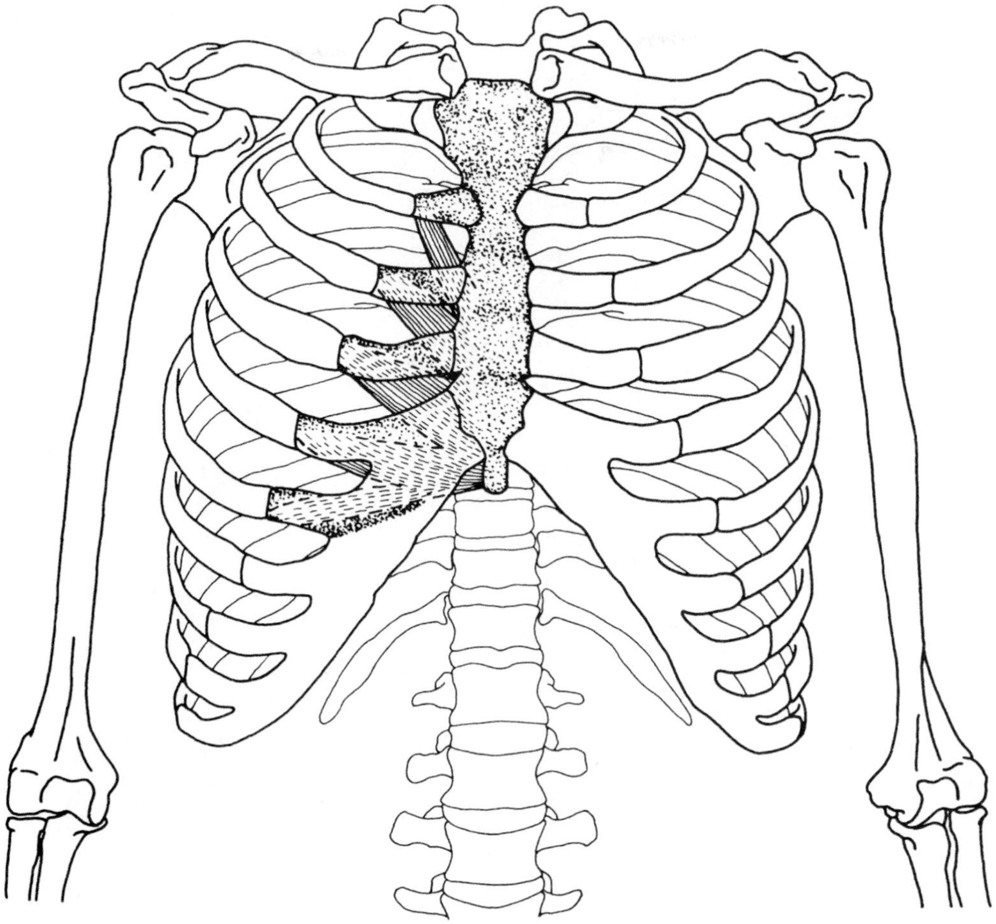

Trunk—anterior view

Origin Inner surface of lower portion of sternum and adjacent costal cartilages

Insertion Inner surfaces of costal cartilages of the second through sixth ribs

Action Draws ventral part of ribs downward, decreasing the volume of the thoracic cavity for forceful expiration

Nerve Intercostal nerves

Note: These muscles are deep to the internal intercostal muscles.

LEVATORES COSTARUM

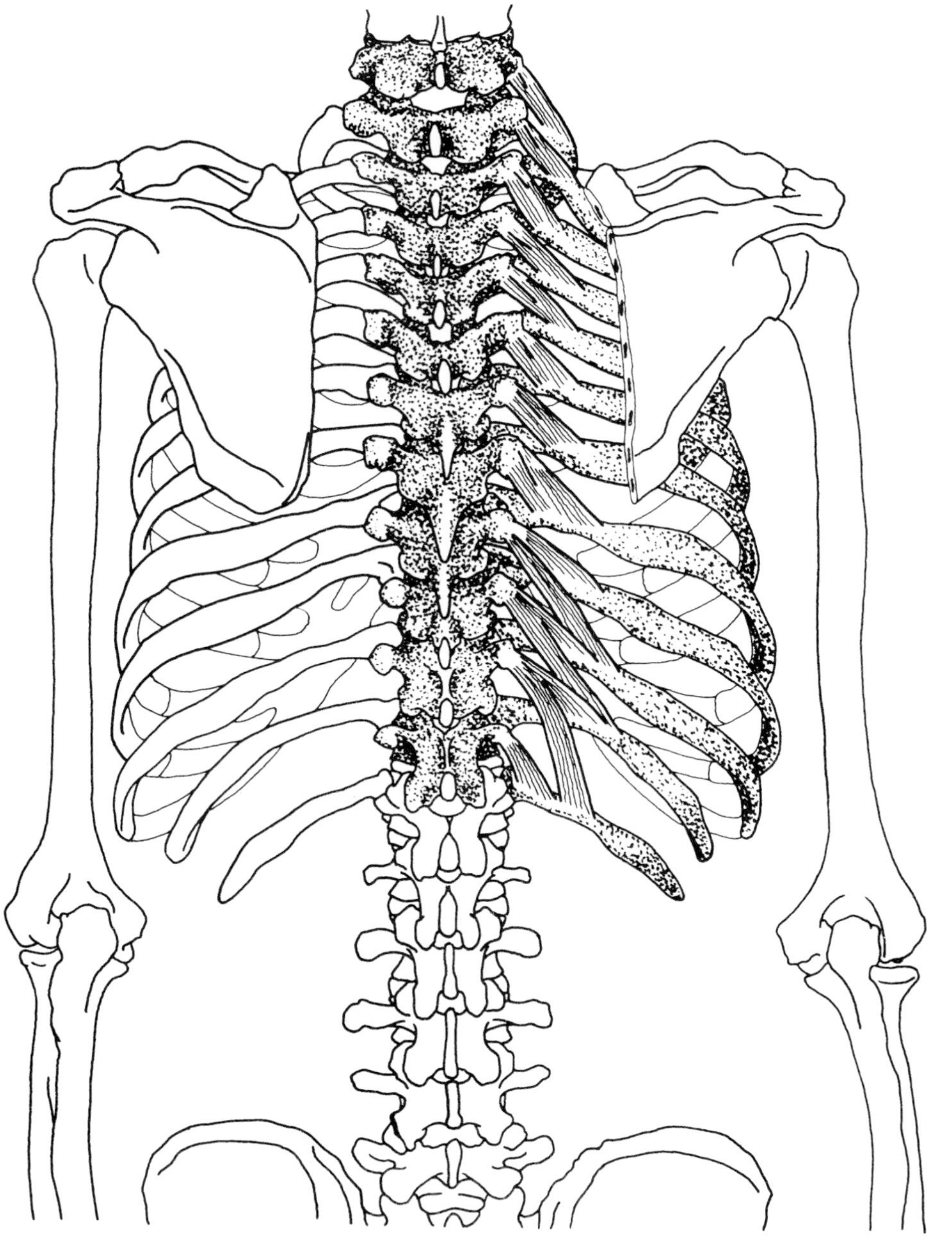

Trunk—dorsal view

Origin Transverse processes of the seventh cervical and the upper eleven thoracic vertebrae

Insertion Laterally to outer surface of next lower rib (lower muscles may cross over one rib)

Action Raises ribs; extends, laterally flexes, and rotates vertebral column

Nerve Intercostal nerves

SERRATUS POSTERIOR SUPERIOR

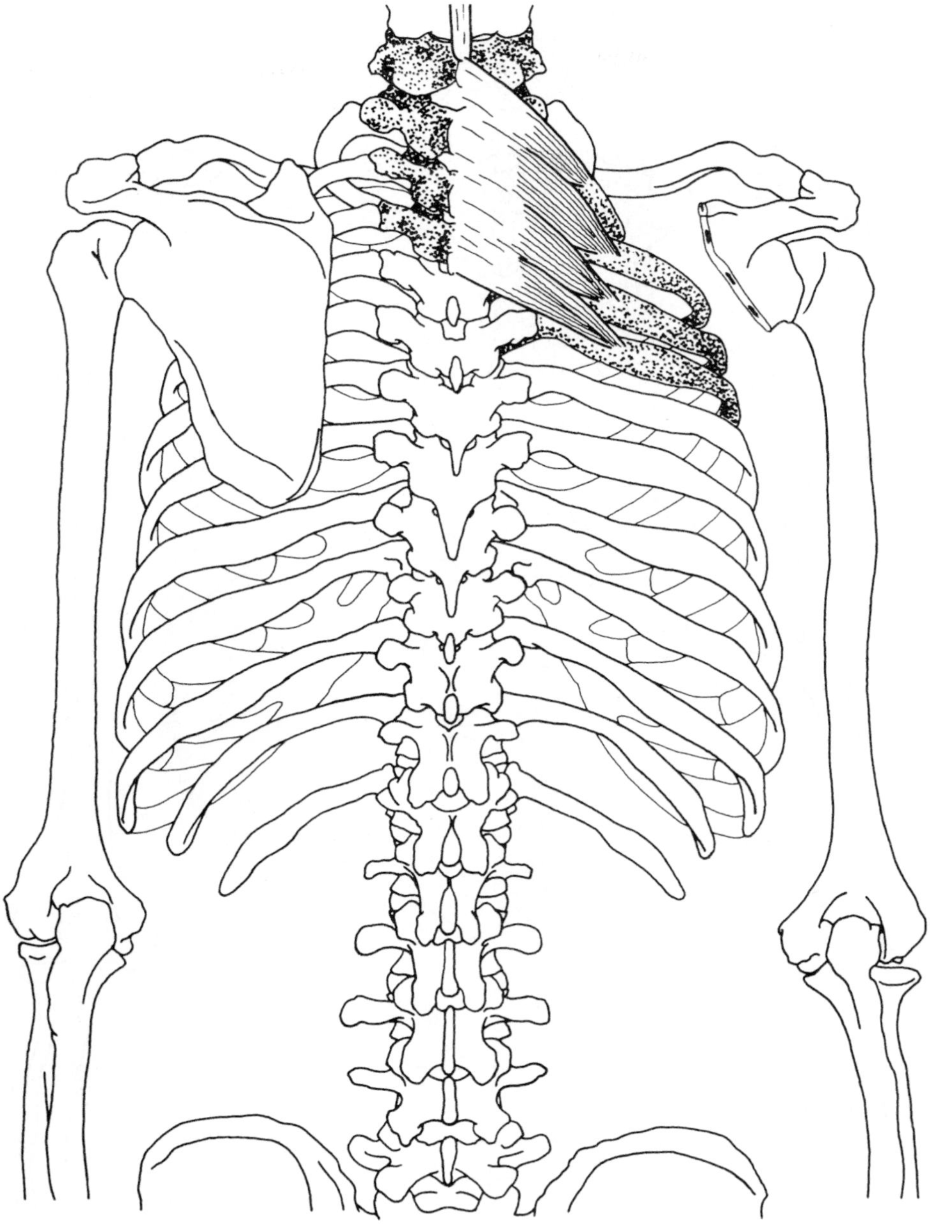

Trunk—dorsal view

Origin Ligamentum nuchae, spinous processes of seventh cervical and first few thoracic vertebrae

Insertion Upper borders of the second through fifth ribs lateral to their angles

Action Raises ribs in inspiration

Nerve T1–T4

SERRATUS POSTERIOR INFERIOR

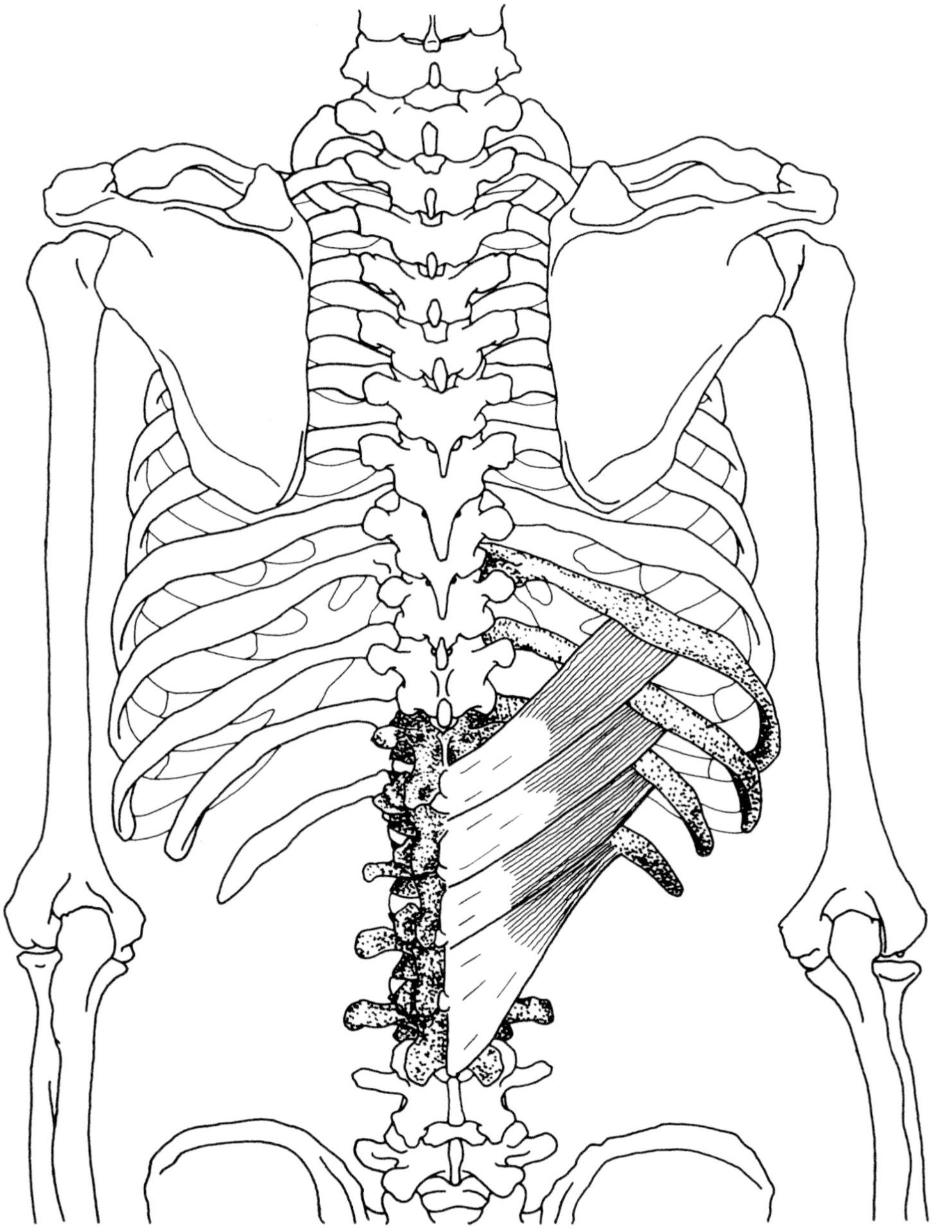

Trunk—dorsal view

Origin Spinous processes of the lower two thoracic and the upper two or three lumbar vertebrae

Insertion Lower borders of bottom four ribs

Action Pulls ribs down, resisting pull of diaphragm

Nerve T9–T12

DIAPHRAGM

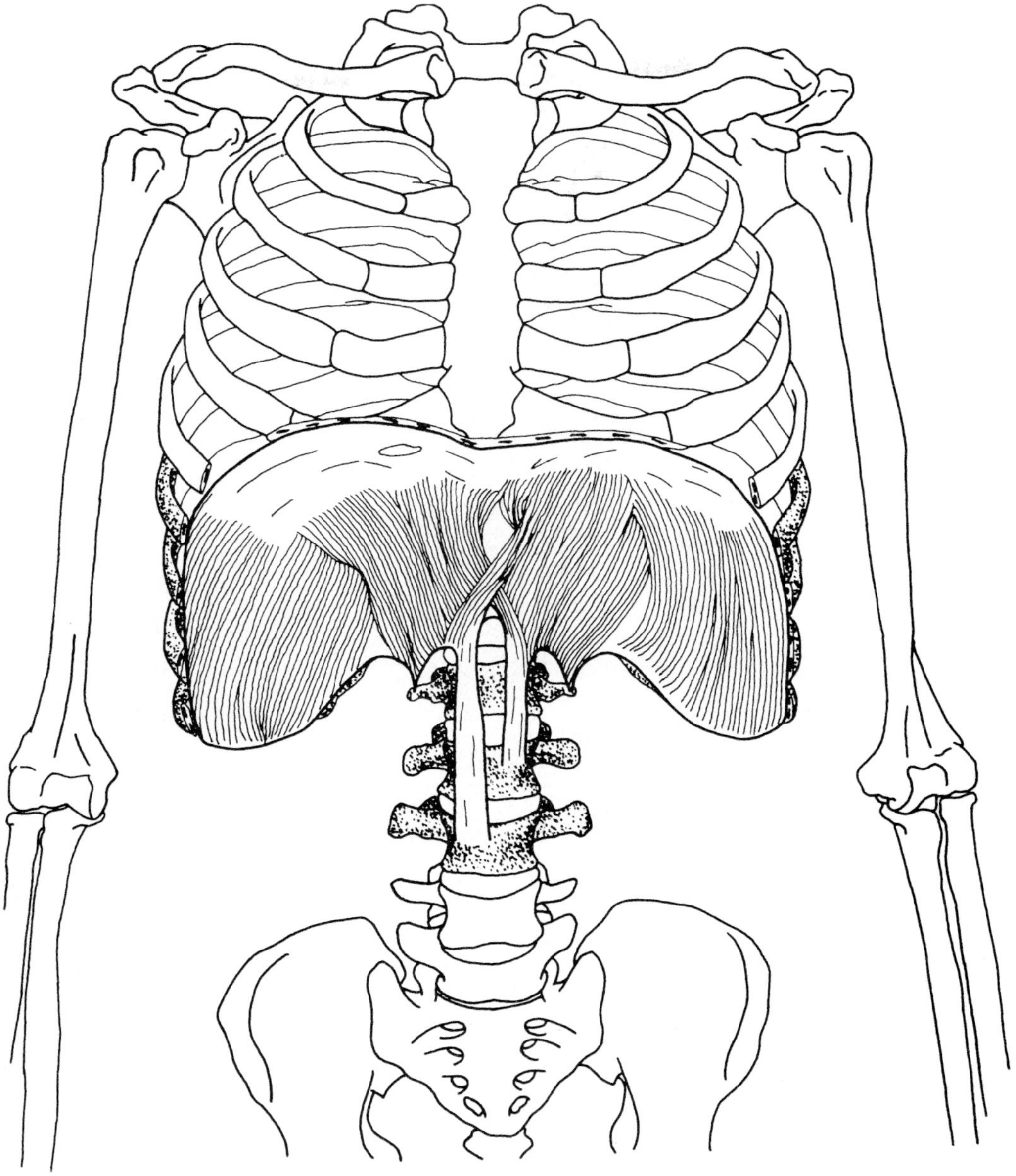

Trunk—anterior view

(Lower costal cartilages removed)

Origin

Sternal part—inner part of xiphoid process

Costal part—inner surfaces of lower six ribs and their cartilages

Lumbar part—upper two or three lumbar vertebrae and lateral and medial lumbocostal arches*

Insertion Fibers converge and meet on a central tendon

Action Draws central tendon inferiorly

Nerve Phrenic nerve (C3–C5)

Note: This muscle inserts upon itself. Its action is to change the volume of the thoracic and abdominal cavities.

*These tendinous structures, also known as the medial and lateral arcuate ligaments, allow the diaphragm to bridge the upper parts of the psoas major and quadratus lumborum muscles.

OBLIQUUS EXTERNUS ABDOMINIS

(External Oblique)

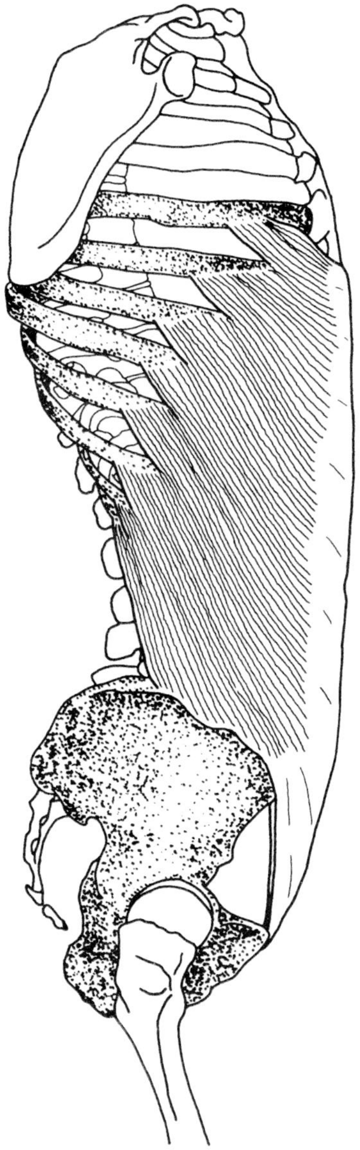

Trunk—lateral view

Origin Lower eight ribs

Insertion Anterior part of iliac crest, abdominal aponeurosis to linea alba

Action Compresses abdominal contents, laterally flexes and rotates vertebral column

Nerve Eighth to twelfth intercostal, iliohypogastric, ilioinguinal nerves

Relationships Most superficial of the three lateral abdominal muscles

Note: Important in forced expiration, coughing, sneezing.

OBLIQUUS INTERNUS ABDOMINIS

(Internal Oblique)

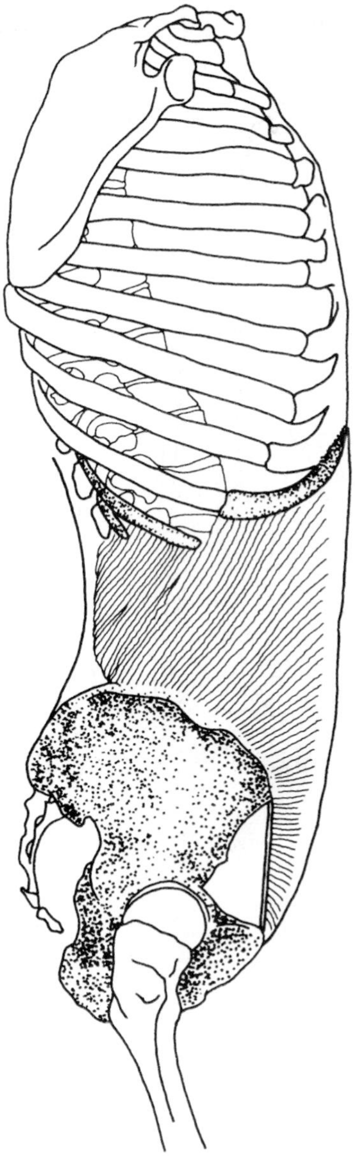

Trunk—lateral view

Origin Lateral half of inguinal ligament, iliac crest, thoracolumbar fascia

Insertion Cartilage of bottom three or four ribs, abdominal aponeurosis to linea alba

Action Compresses abdominal contents, laterally flexes and rotates vertebral column

Nerve Eighth to twelfth intercostal, iliohypogastric, ilioinguinal nerves

Relationships Middle layer of the three lateral abdominal muscles

Note: Important in forced expiration, coughing, sneezing.

CREMASTER

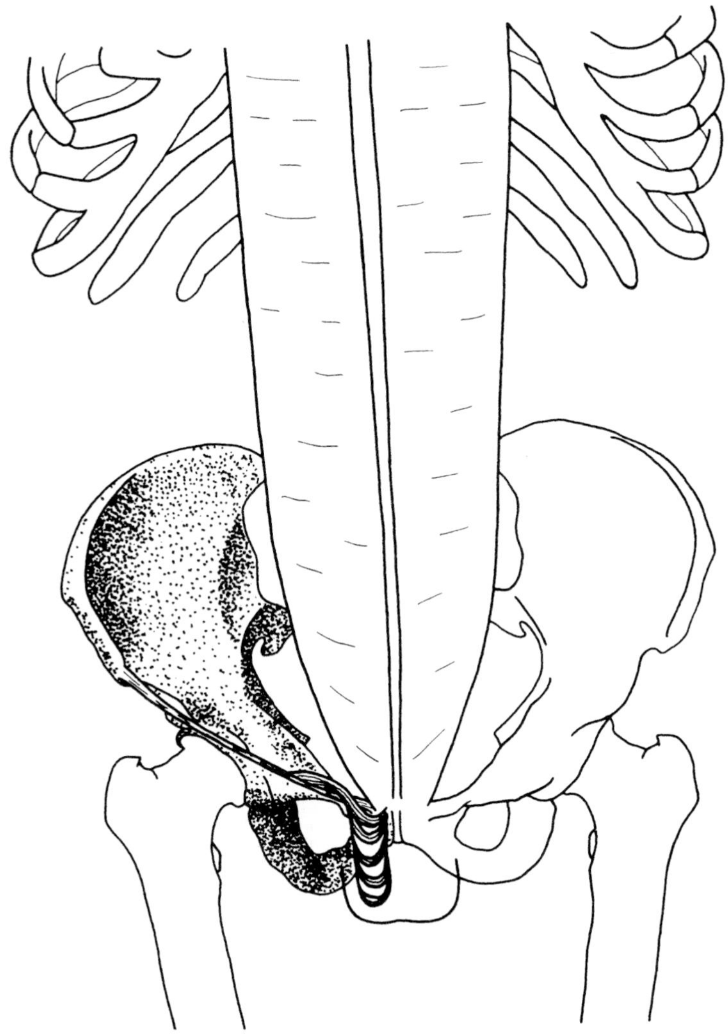

Trunk—anterior view

Origin Inguinal ligament

Insertion Pubic tubercle, crest of pubis, sheath of rectus abdominis

Action Pulls testes toward body

Nerve Genital branch of genitofemoral nerve

Note: The cremaster regulates the temperature of the testes, which is important for spermatogenesis.

TRANSVERSUS ABDOMINIS

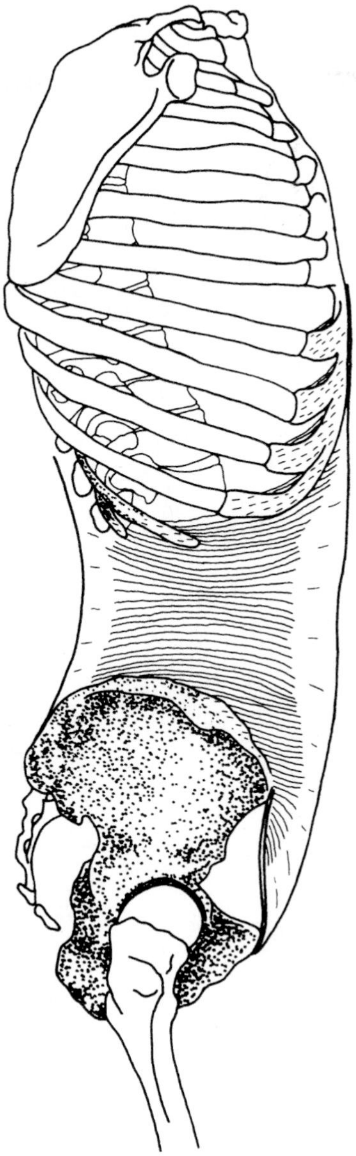

Trunk—lateral view

Origin Lateral part of inguinal ligament, iliac crest, thoracolumbar fascia, cartilage of lower six ribs

Insertion Abdominal aponeurosis to linea alba

Action Compresses abdomen

Nerve Seventh to twelfth intercostal, iliohypogastric, ilioinguinal nerves

Relationships Deepest of the three lateral abdominal muscles

Note: Important in forced expiration, coughing, sneezing.

RECTUS ABDOMINIS*

Origin	Crest of pubis, pubic symphysis
Insertion	Cartilage of fifth, sixth, and seventh ribs, xiphoid process
Action	Flexes vertebral column, compresses abdomen
Nerve	Seventh through twelfth intercostal nerves

*Tendinous bands divide each rectus into three or four bellies. Each rectus is sheathed in aponeurotic fibers from the lateral abdominal muscles. These fibers meet centrally to form the linea alba.

Note: The pyramidalis is a small, unimportant muscle that extends from the ventral surface of the pubis to the lower part of the linea alba. It is frequently absent.

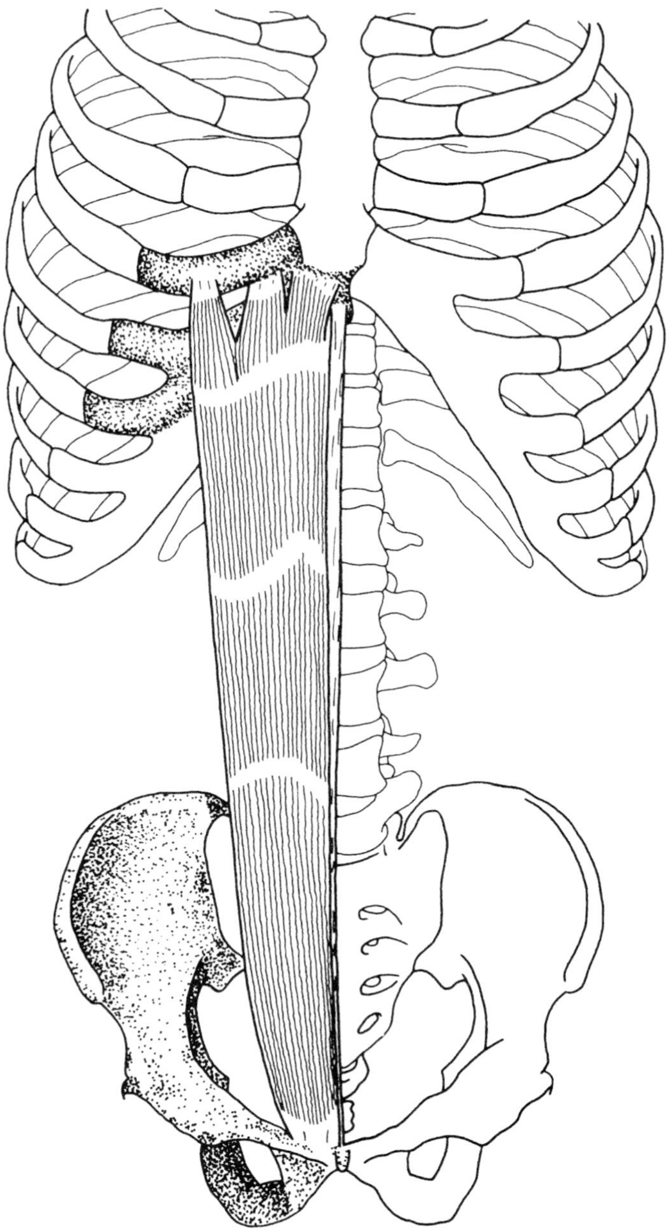

Trunk—anterior view

ABDOMINAL MUSCLES

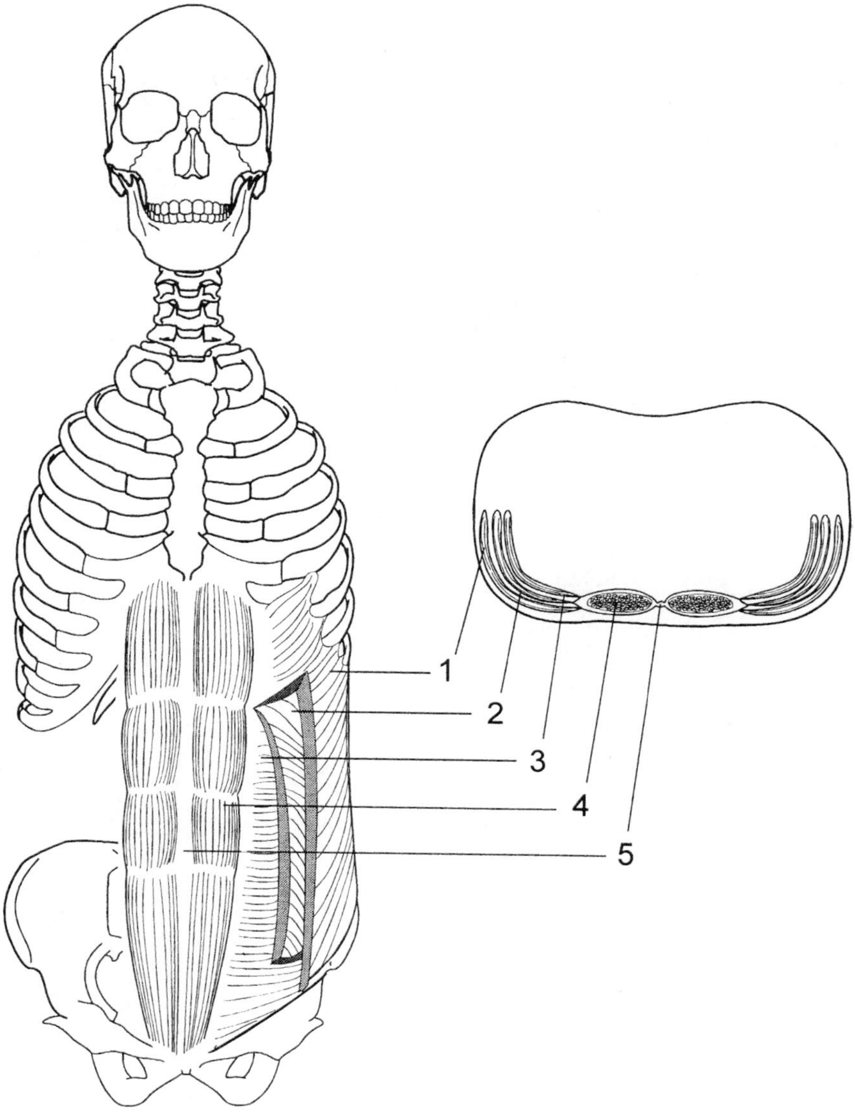

Trunk—anterior and cross-sectional views

1. Obliquus externus abdominis
2. Obliquus internus abdominis
3. Transversus abdominis
4. Rectus abdominis
5. Linea alba

Note: The aponeuroses (tendons) of the three lateral abdominal muscles join to form the fascial sheath surrounding the rectus abdominis.

QUADRATUS LUMBORUM

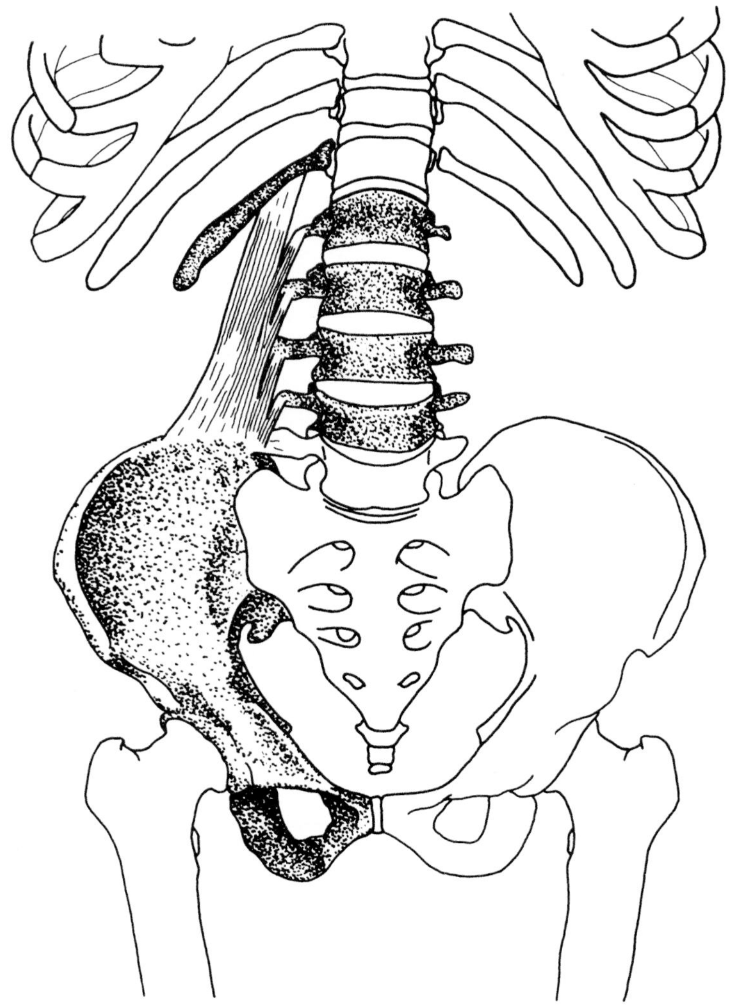

Lower trunk—anterior view

Origin Iliolumbar ligament, iliac crest

Insertion Twelfth rib, transverse processes of upper four lumbar vertebrae

Action Laterally flexes vertebral column, fixes ribs for forced expiration

Nerve T12, L1

Note: Fixation of the ribs may provide a stable attachment of the diaphragm for voice control in singers.

CHAPTER SIX
MUSCLES OF THE SHOULDER AND ARM

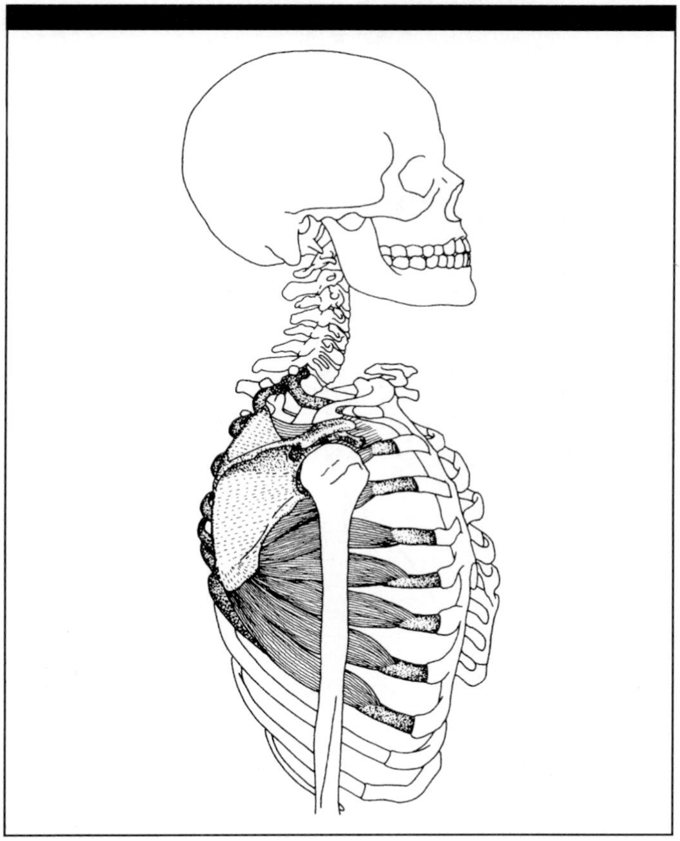

PECTORALIS MAJOR

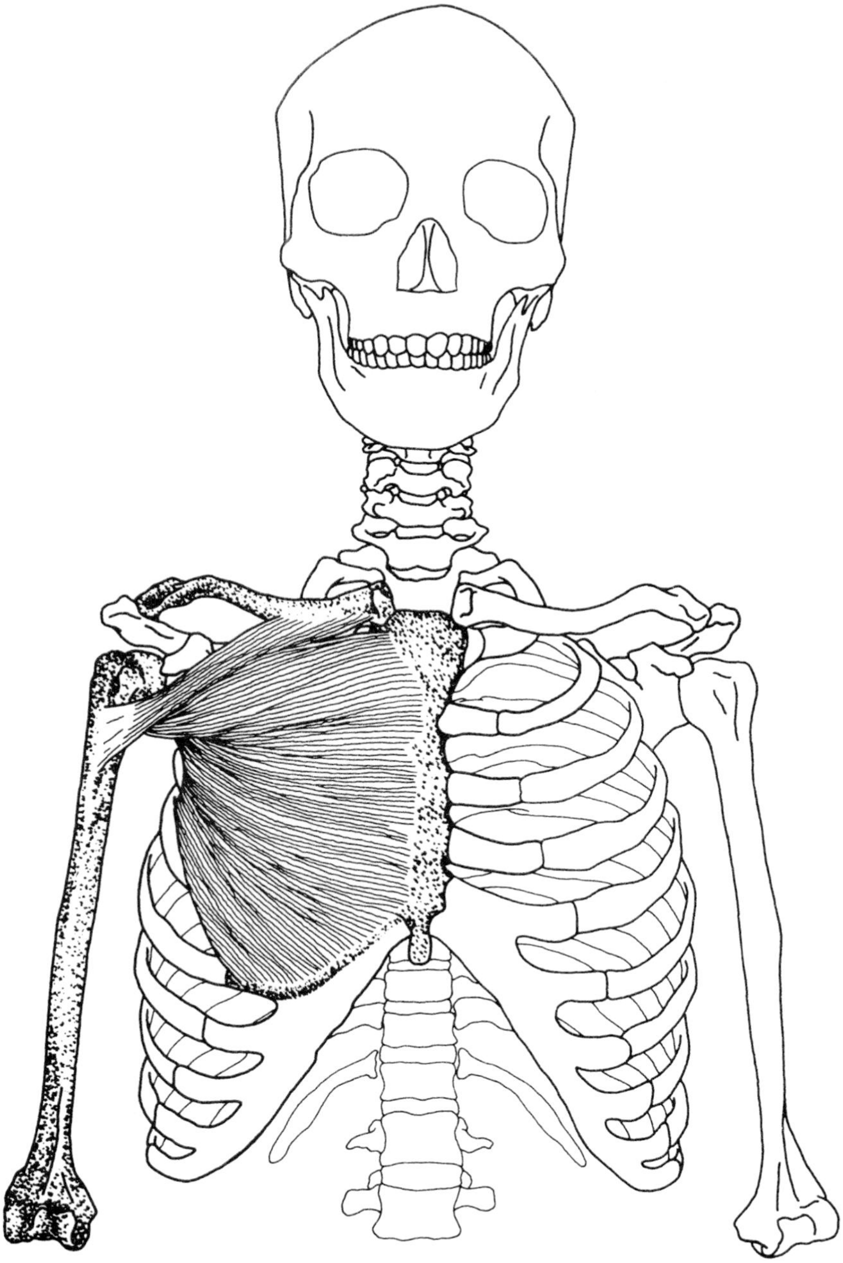

Anterior view

Origin Clavicular part—medial half of the clavicle
Sternocostal part—sternum, upper six costal cartilages, aponeurosis of external oblique

Insertion Lateral lip of intertubercular (bicipital) groove of humerus, crest below greater tubercle of the humerus

Action Both parts adduct, medially rotate arm; clavicular part flexes arm from full extension; sternocostal part extends the flexed arm

Nerve Medial and lateral pectoral nerves (C5–C8, T1)

PECTORALIS MINOR

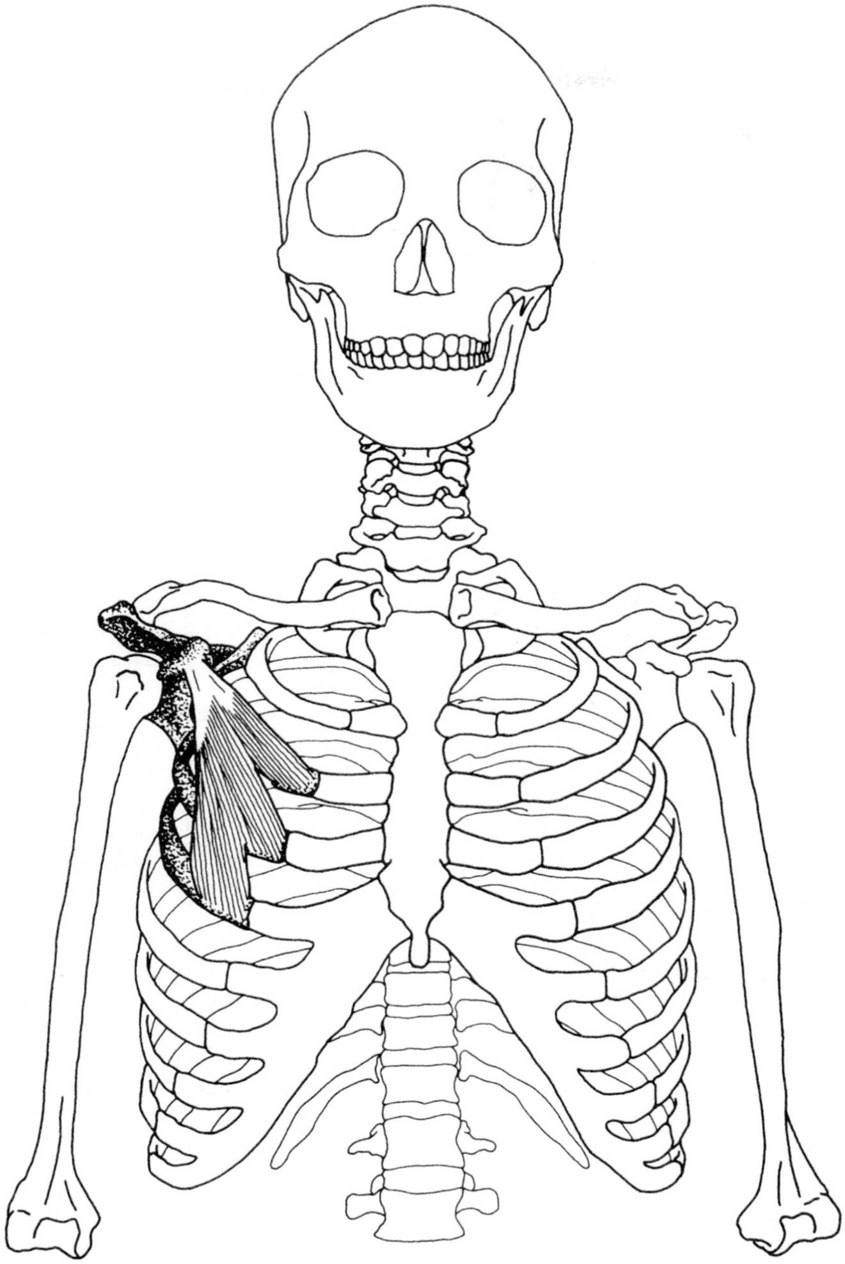

Anterior view

Origin External surfaces of the third, fourth, and fifth ribs

Insertion Coracoid process of the scapula

Action Draws scapula forward and downward, raises ribs* in forced inspiration

Nerve Medial pectoral nerve (C8, T1)

Relationships Deep to pectoralis major

*Raising the ribs requires stabilization of the scapula by the rhomboids and trapezius.

SUBCLAVIUS

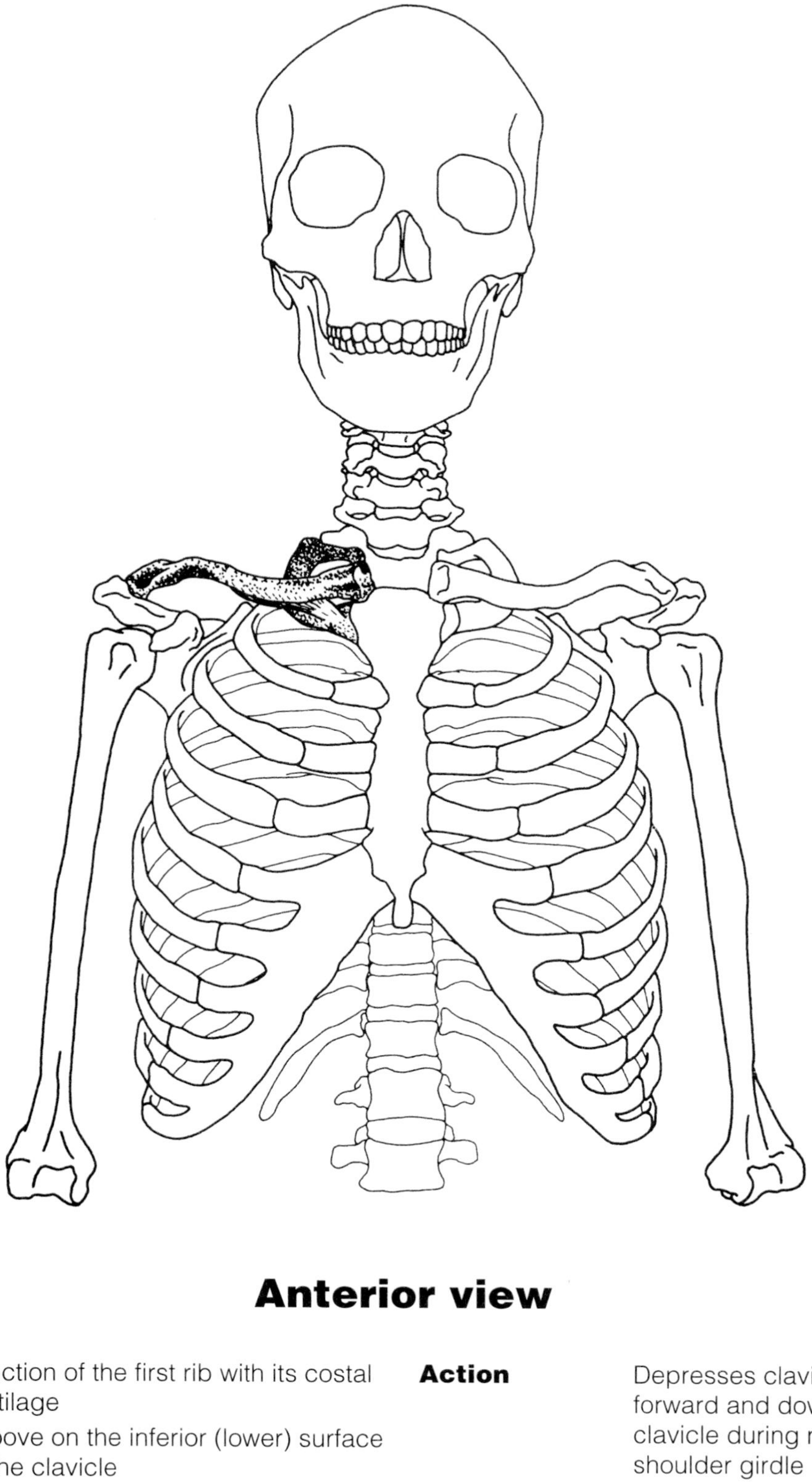

Anterior view

Origin Junction of the first rib with its costal cartilage

Insertion Groove on the inferior (lower) surface of the clavicle

Action Depresses clavicle, draws shoulder forward and downward, steadies clavicle during movements of shoulder girdle

Nerve C5, C6

CORACOBRACHIALIS

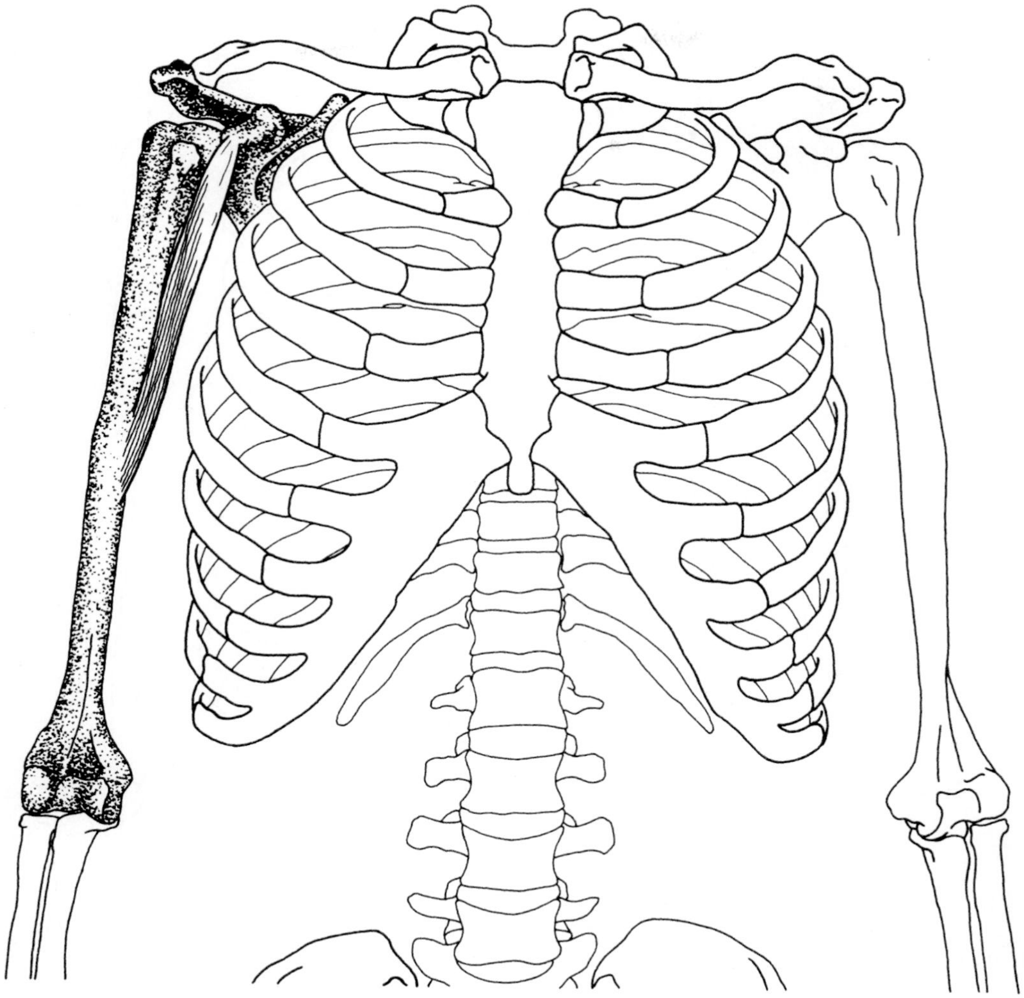

Anterior view

Origin Tip (apex) of the coracoid process of scapula

Insertion Middle third of the medial surface and border of the humerus

Action Weakly adducts arm (flexion unsubstantiated), aids in stabilizing humerus

Nerve Musculocutaneous nerve (C6, C7)

Relationships Deep to short head of biceps

BICEPS BRACHII

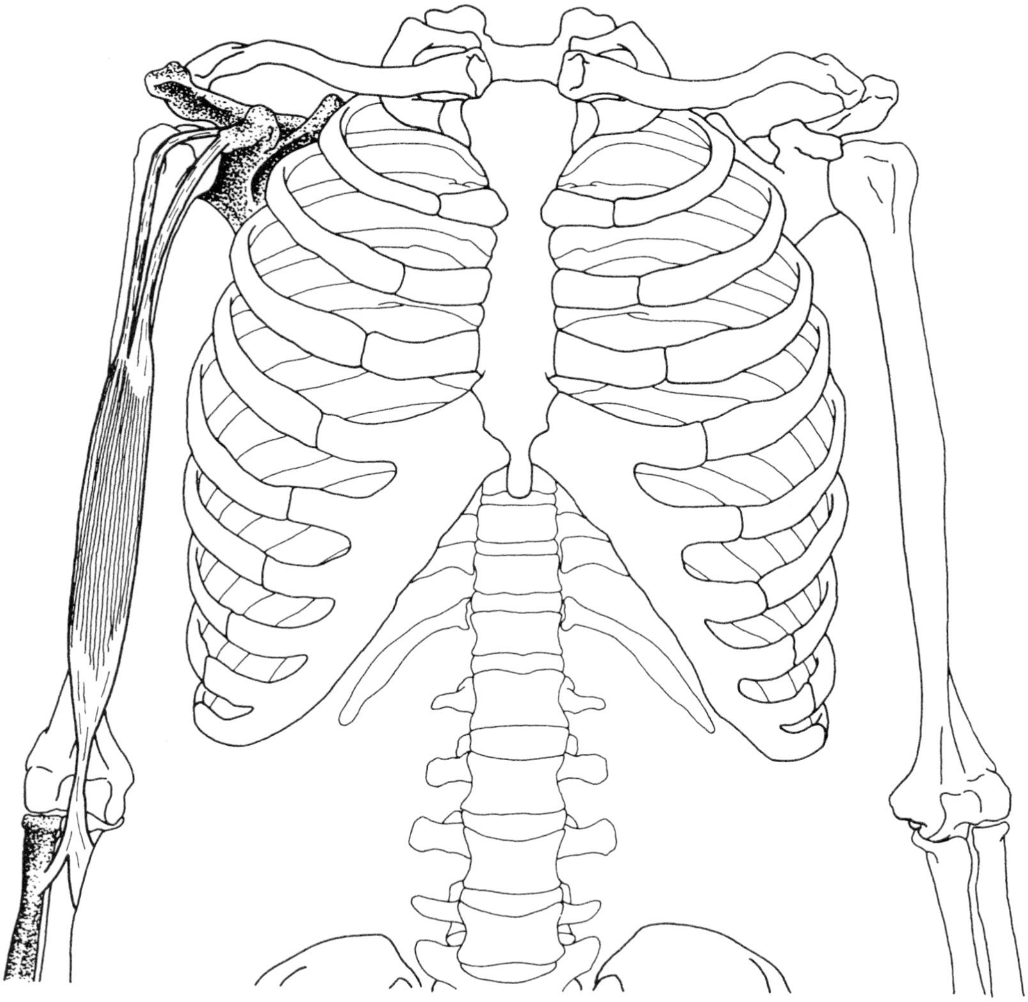

Anterior view

Origin Long head—supraglenoid tubercle of scapula

Short head—coracoid process of scapula

Insertion Tuberosity of radius, bicipital aponeurosis into deep fascia on medial part of forearm

Action Supinates forearm, flexes forearm, weakly flexes arm at shoulder

Nerve Musculocutaneous nerve (C5, C6)

Relationships Long head passes through intertubercular (bicipital) groove, then inside glenohumeral joint capsule

BRACHIALIS

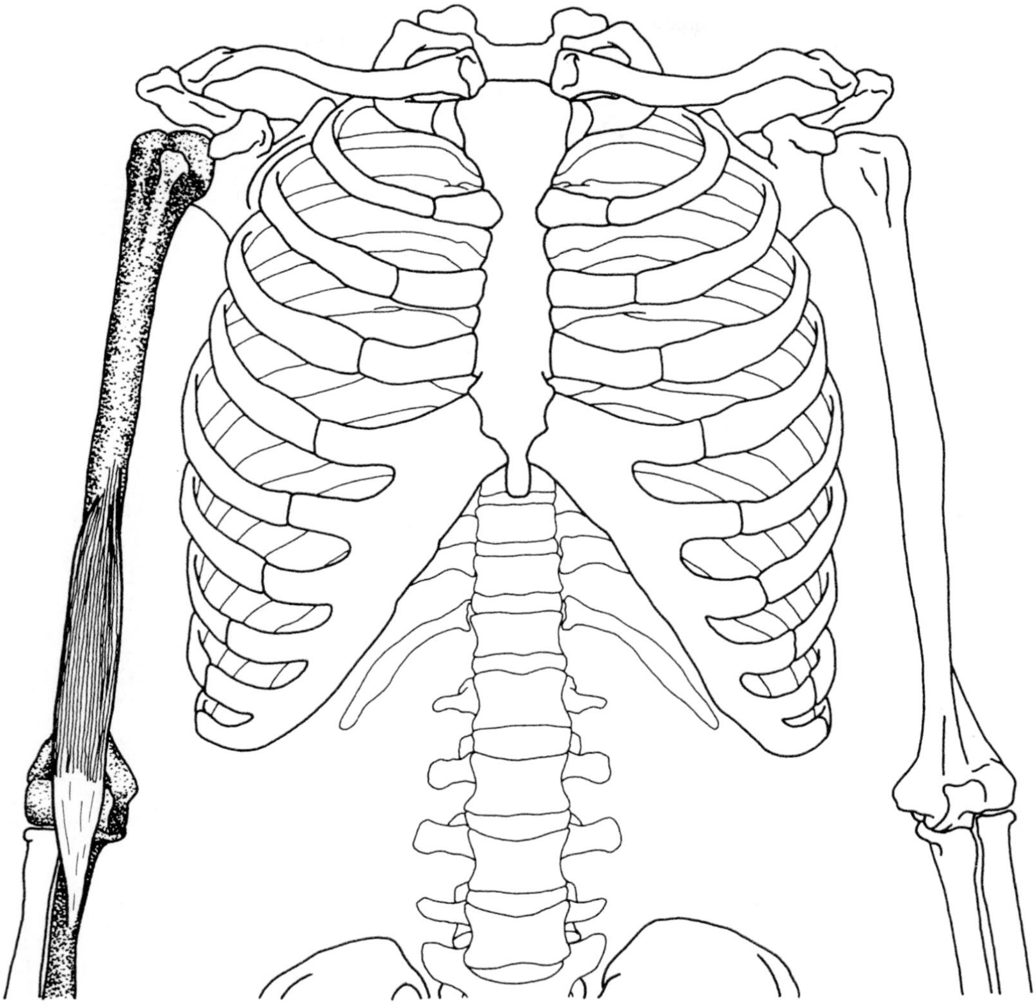

Anterior view

Origin	Anterior of lower half of humerus	**Action**	Flexes forearm
Insertion	Coronoid process of ulna, tuberosity of ulna	**Nerve**	Musculocutaneous nerve (C5, C6)
		Relationships	Deep to biceps brachii

MUSCLES OF THE ANTERIOR CHEST AND ARM

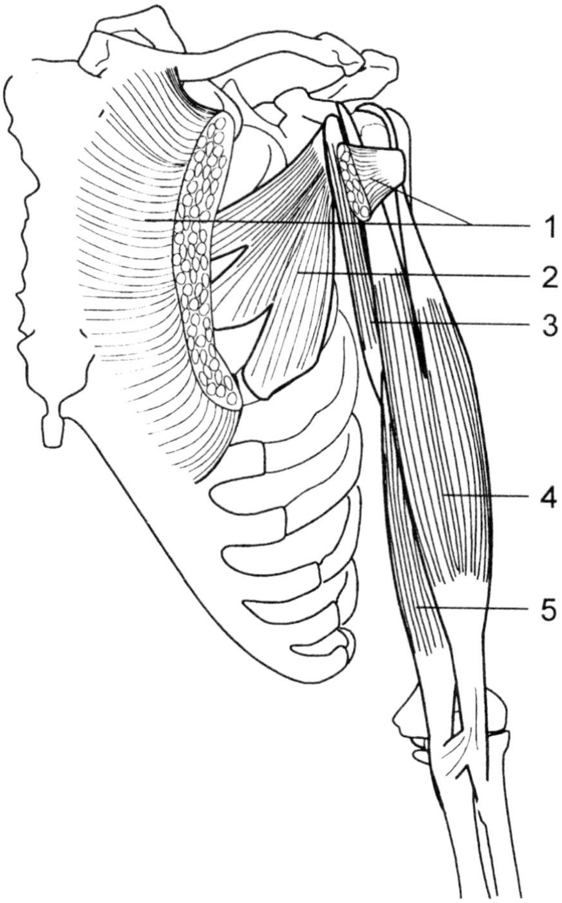

Shoulder—anterior view

1. Pectoralis major (cut)
2. Pectoralis minor
3. Coracobrachialis
4. Biceps brachii
5. Brachialis

TRAPEZIUS

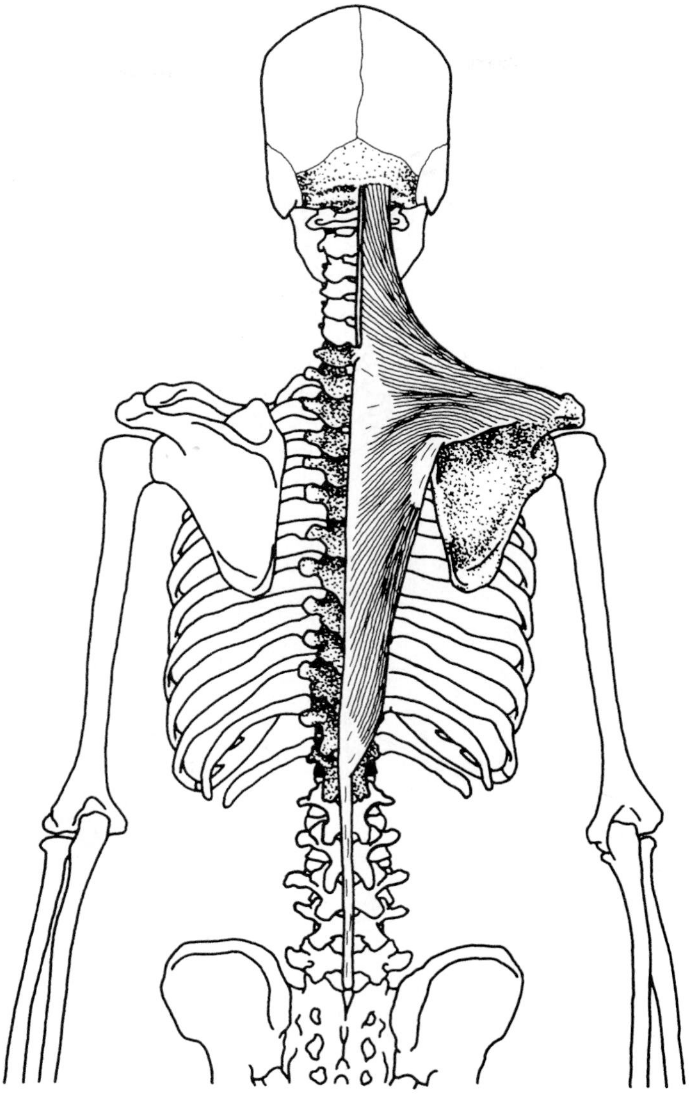

Posterior view

Origin Medial third of superior nuchal line, external occipital protuberance, ligamentum nuchae, spinous processes and supraspinous ligaments of seventh cervical and all thoracic vertebrae

Insertion Upper part—lateral third of clavicle

Middle part—acromion and crest of spine of scapula

Lower part—medial portion of crest of spine of scapula (tubercle)

Action Upper part elevates scapula,* middle part retracts (adducts) scapula, lower part depresses scapula, upper and lower parts together rotate scapula (important in elevating arm)

Nerve Accessory (eleventh cranial), C3, C4

Relationships Most superficial muscle of back

*Upper part stabilizes scapula against downward rotation, as when weight is carried in the hand.

LATISSIMUS DORSI

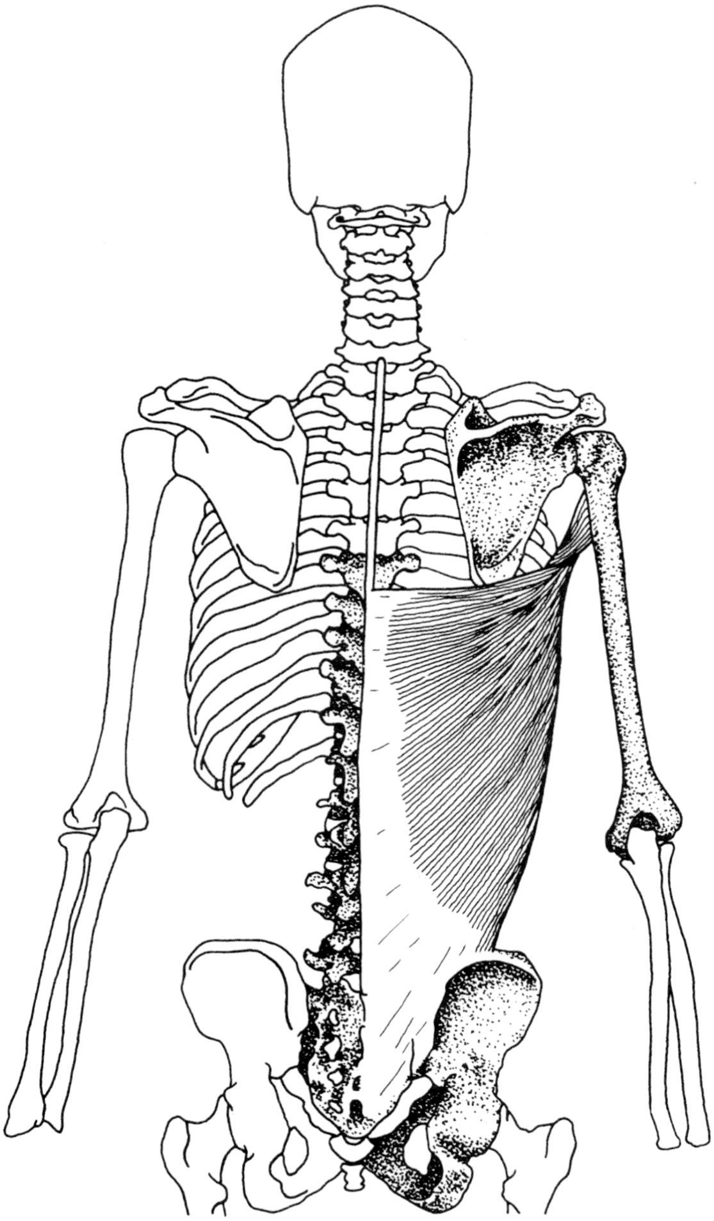

Posterior view

Origin Spinous processes of the lower six thoracic vertebrae, lumbar vertebrae, sacral vertebrae, supraspinal ligament, and posterior part of the iliac crest through the lumbar (thoracolumbar) fascia, lower three or four ribs, inferior angle of the scapula

Insertion Floor (bottom) of the bicipital groove of humerus

Action Extends, adducts, and medially rotates the arm, draws the shoulder downward and backward, keeps inferior angle of scapula against the chest wall, accessory muscle of respiration

Nerve Thoracodorsal nerve (C6–C8)

Note: This muscle is used for the crawl stroke in swimming.

LEVATOR SCAPULAE

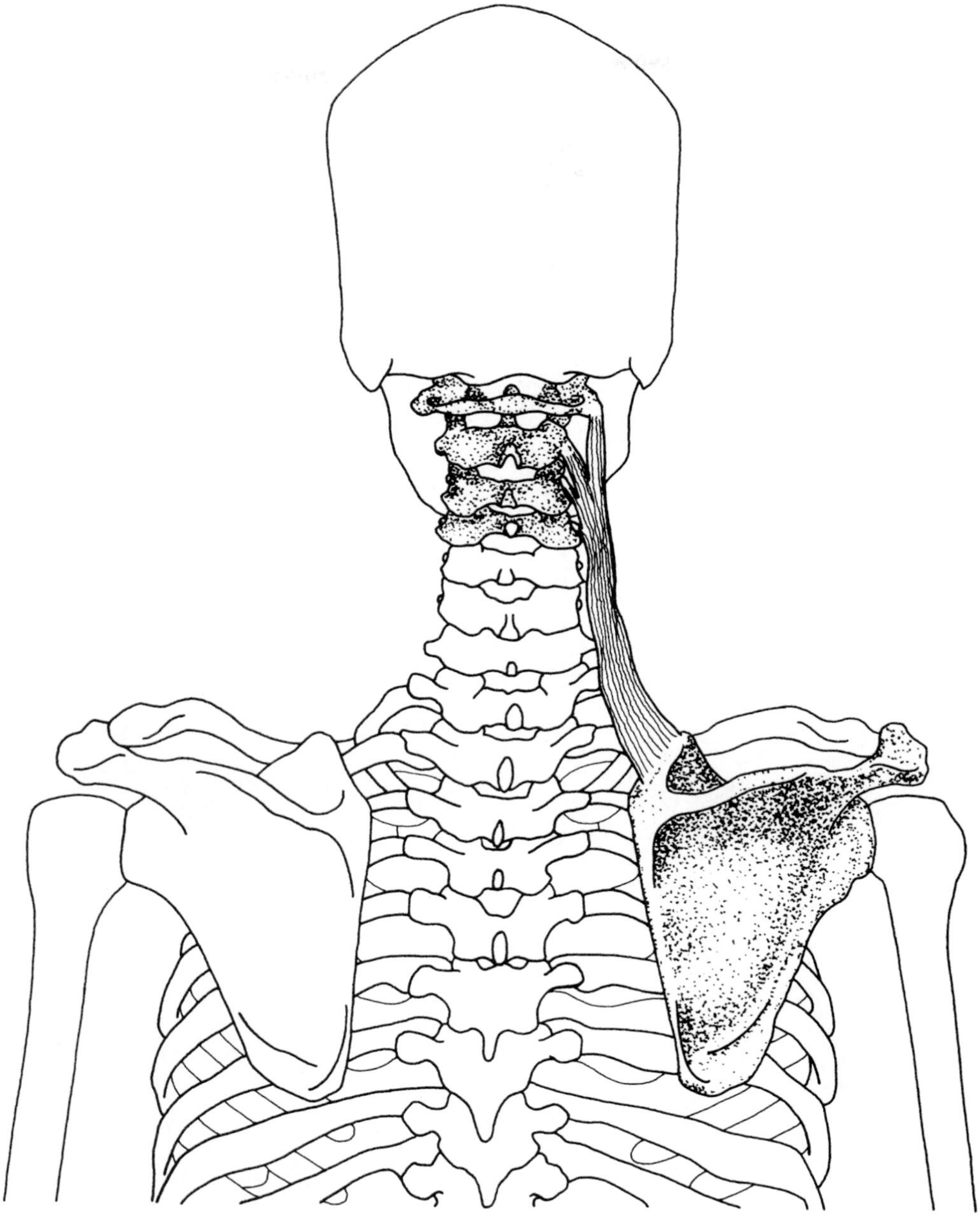

Posterior view

Origin Posterior tubercles of the transverse processes of the first four cervical vertebrae

Insertion Vertebral (medial) border of the scapula at and above the spine

Action Elevates medial border of scapula, rotates scapula to lower the lateral angle, acts with trapezius and rhomboids to pull scapula medially and upward, bends neck laterally

Nerve Dorsal scapular nerve (C5)

RHOMBOIDEUS MAJOR

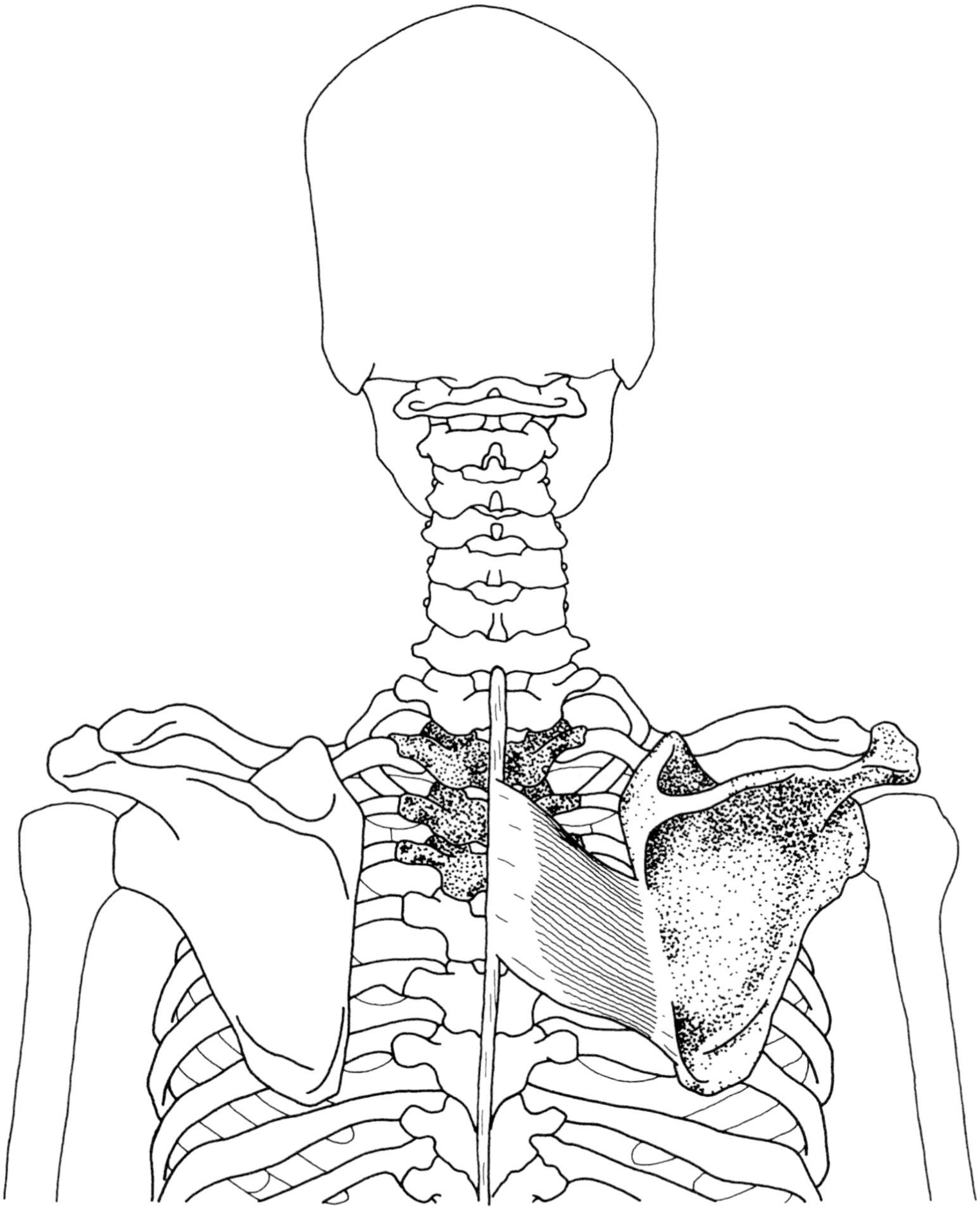

Posterior view

Origin Spines of the second to fifth thoracic vertebrae, supraspinous ligament

Insertion Medial border of the scapula below the spine

Action Retracts and stabilizes scapula, elevates the medial border of the scapula causing downward rotation, assists in adduction of arm

Nerve Dorsal scapular nerve (C5)

RHOMBOIDEUS MINOR

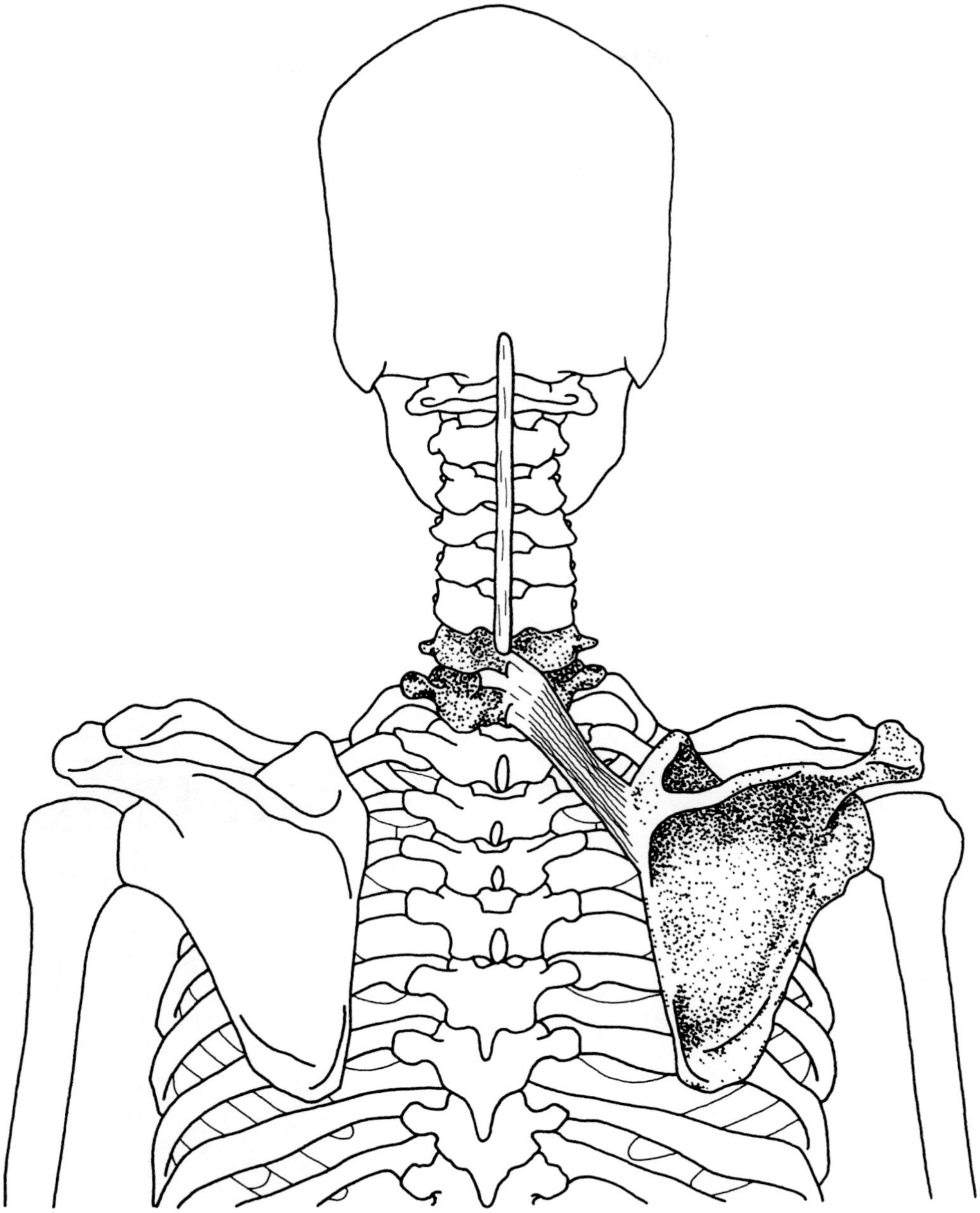

Posterior view

Origin Spines of the seventh cervical and first thoracic vertebrae, lower part of the ligamentum nuchae

Insertion Medial border of the scapula at the root of the spine

Action Retracts and stabilizes scapula, elevates the medial border of the scapula, rotates the scapula to depress the lateral angle (assists in adduction of arm)

Nerve Dorsal scapular nerve (C5)

SERRATUS ANTERIOR

Origin Outer surfaces and superior borders of first eight or nine ribs, and fascia covering first intercostal space

Insertion Anterior surface (costal surface) of the medial border of the scapula

Action Rotates scapula for abduction and flexion of arm, protracts scapula

Nerve Long thoracic nerve (C5–C7)

Relationships Serratus anterior and rhomboids both insert on the medial border of scapula; they are antagonists causing protraction and retraction

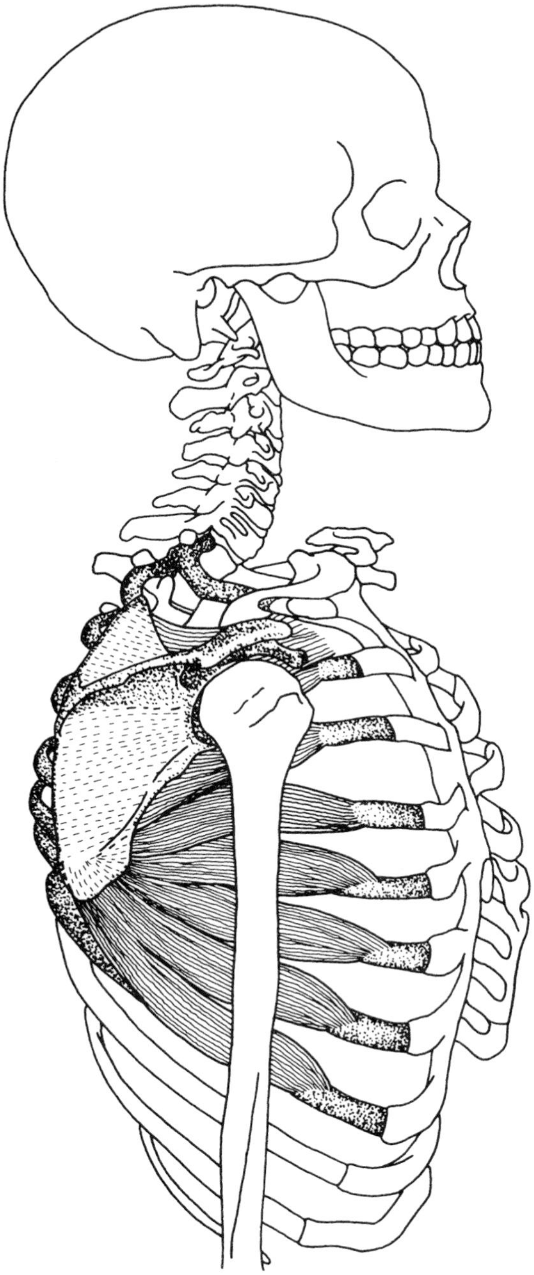

Lateral view

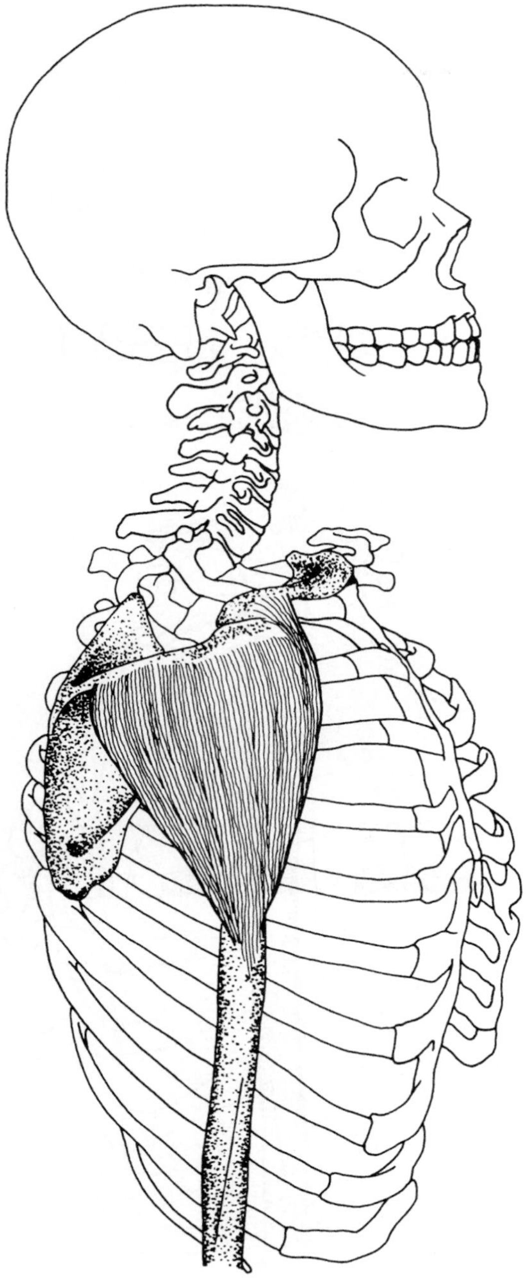

Lateral view

DELTOIDEUS

Origin
Anterior portion—anterior border and superior surface of the lateral third of the clavicle

Middle portion—lateral border of the acromion process

Posterior portion—lower border of the crest of the spine of the scapula

Insertion
Deltoid tuberosity, on the middle of the lateral surface of the shaft of the humerus

Action
Anterior portion—flexes and medially rotates arm

Middle portion—abducts arm

Posterior portion—extends and laterally rotates arm

Nerve
Axillary nerve (C5, C6)

SUPRASPINATUS

(Rotator cuff)*

Origin	Supraspinous fossa of scapula
Insertion	Upper part of the greater tuberosity of the humerus, capsule of the shoulder joint
Action	Aids deltoid in abduction of arm, draws humerus toward glenoid fossa preventing deltoid from forcing humerus up against acromion, weakly flexes arm
Nerve	Suprascapular nerve (C5)

*Supraspinatus, infraspinatus, teres minor, and subscapularis together are called the rotator cuff. They prevent the larger muscles from dislocating the humerus during their actions.

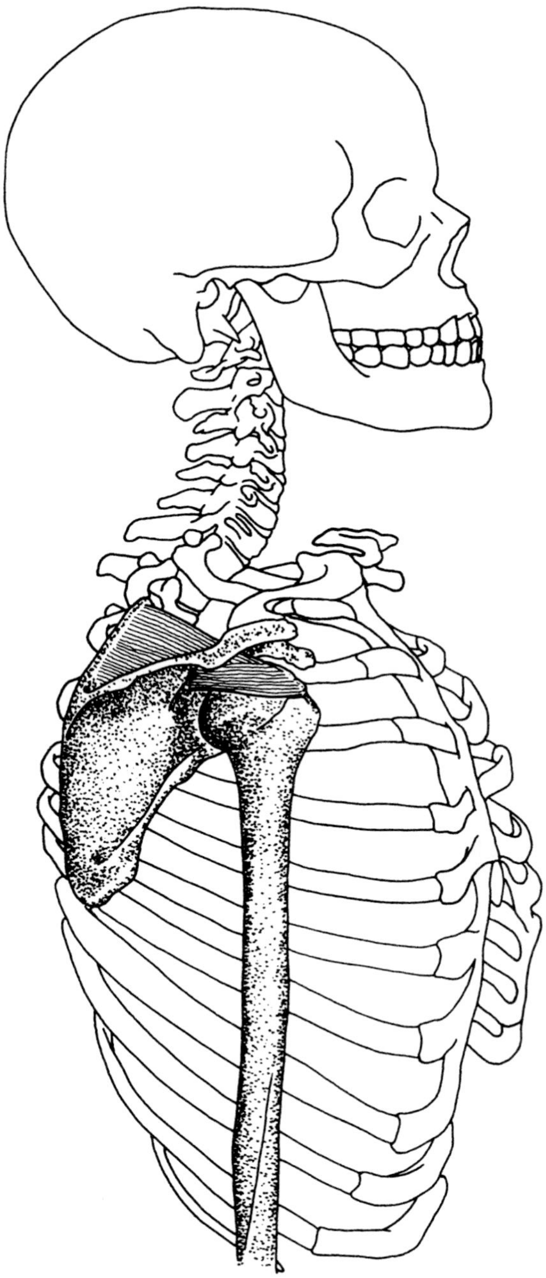

Lateral view

INFRASPINATUS

(Rotator cuff)

Origin	Infraspinous fossa of the scapula
Insertion	Middle facet of the greater tuberosity of the humerus, capsule of the shoulder joint
Action	Draws humerus toward glenoid fossa thus resisting posterior dislocation of arm, as in crawling; laterally rotates; abducts arm
Nerve	Suprascapular nerve (C5, C6)

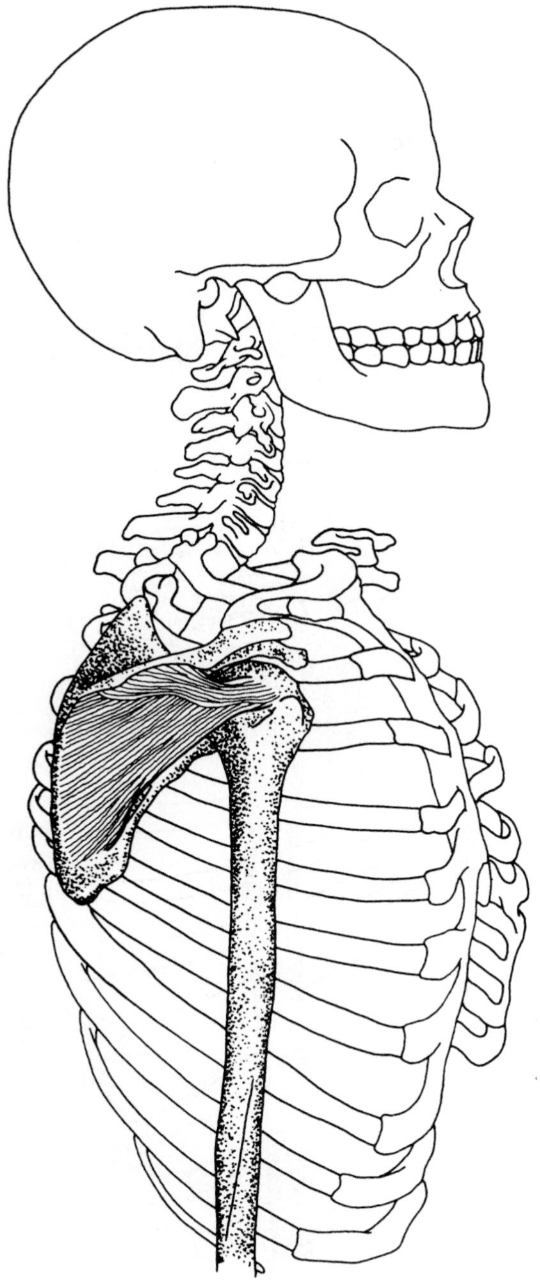

Lateral view

TERES MINOR

(Rotator cuff)

Origin Upper two-thirds of the dorsal surface of the axillary border of the scapula

Insertion The capsule of the shoulder joint, the lower facet of the greater tuberosity of the humerus

Action Laterally rotates arm, weakly adducts arm, draws humerus toward glenoid fossa

Nerve Axillary nerve (C5)

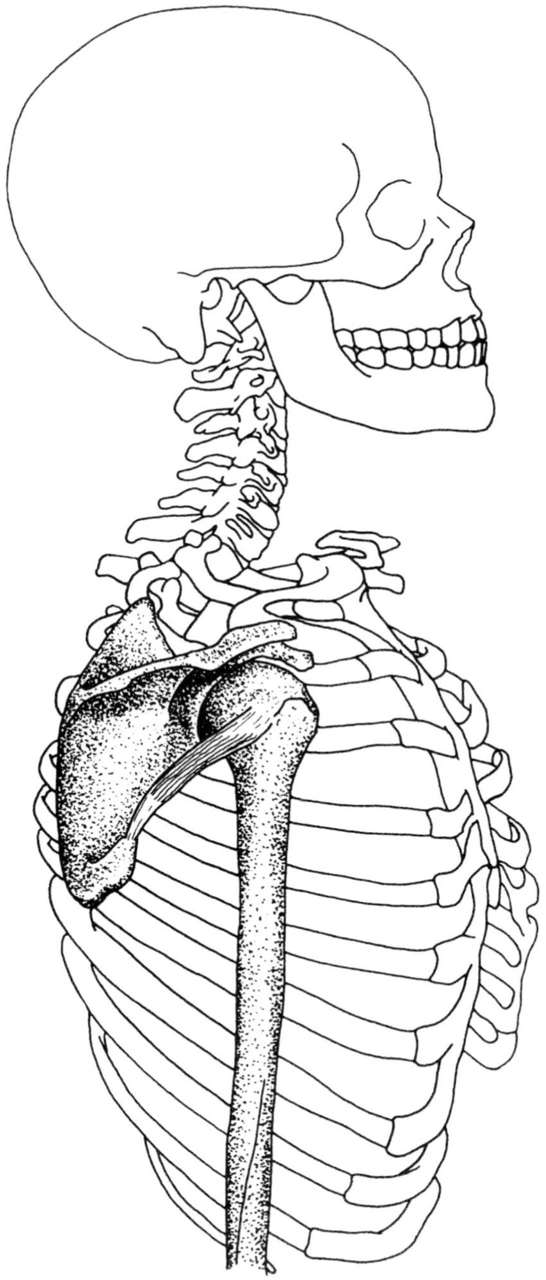

Lateral view

SUBSCAPULARIS

(Rotator cuff)

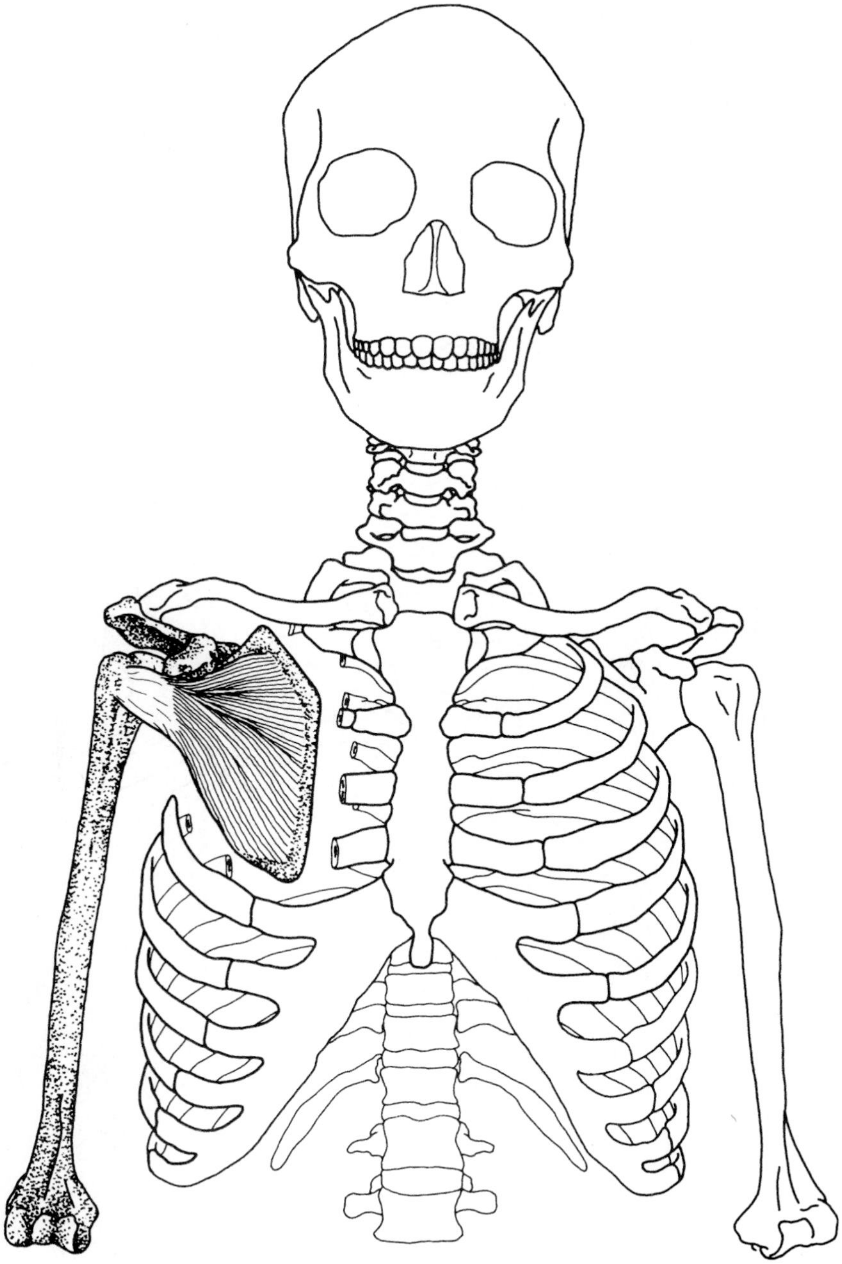

Anterior view

(Upper ribs cut away)

Origin Subscapular fossa on the anterior surface of scapula

Insertion Lesser tuberosity of the humerus, ventral part of the capsule of the shoulder joint

Action Medially rotates arm, stabilizes glenohumeral joint

Nerve Upper and lower subscapular nerves (C5, C6)

TERES MAJOR

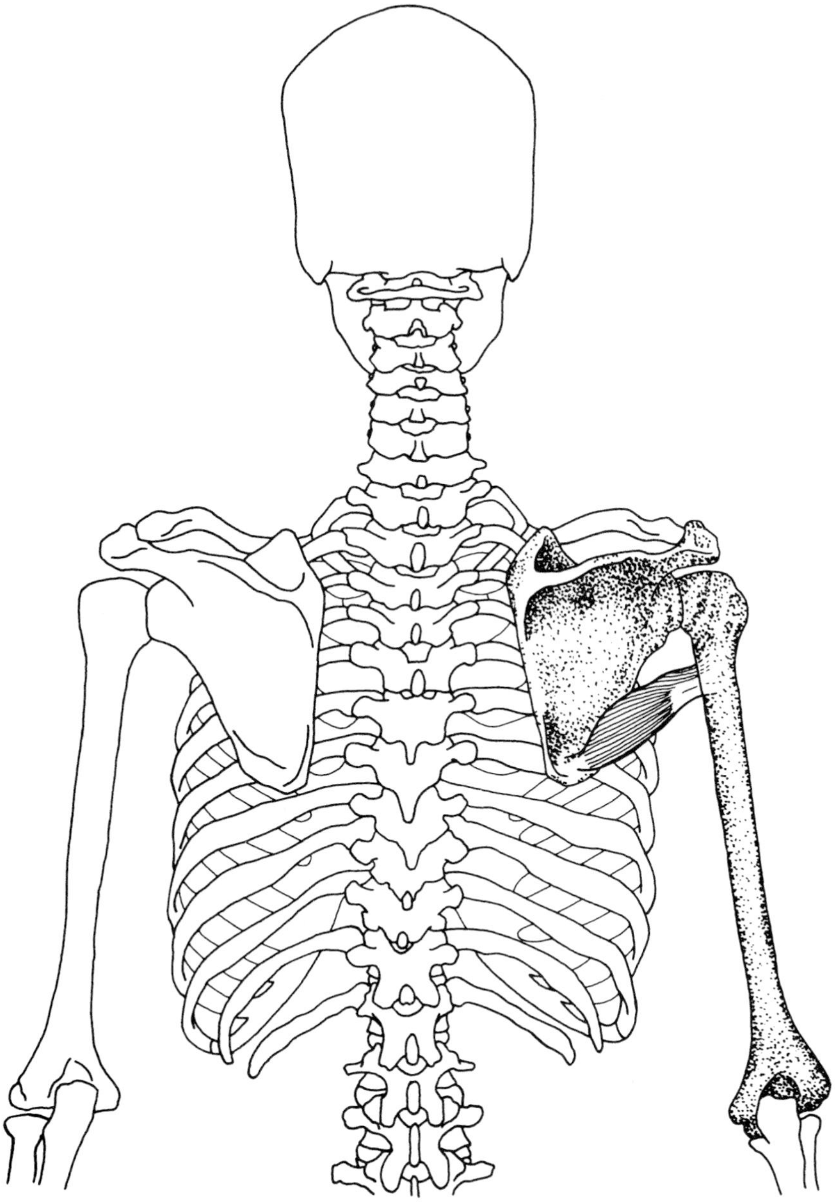

Posterior view

Origin Lower third of the posterior surface of the lateral border of the scapula, near the inferior angle

Insertion Medial lip of the bicipital groove of the humerus

Action Medially rotates arm, adducts arm, extends arm

Nerve Lower subscapular nerve (C5, C6)

TRICEPS BRACHII

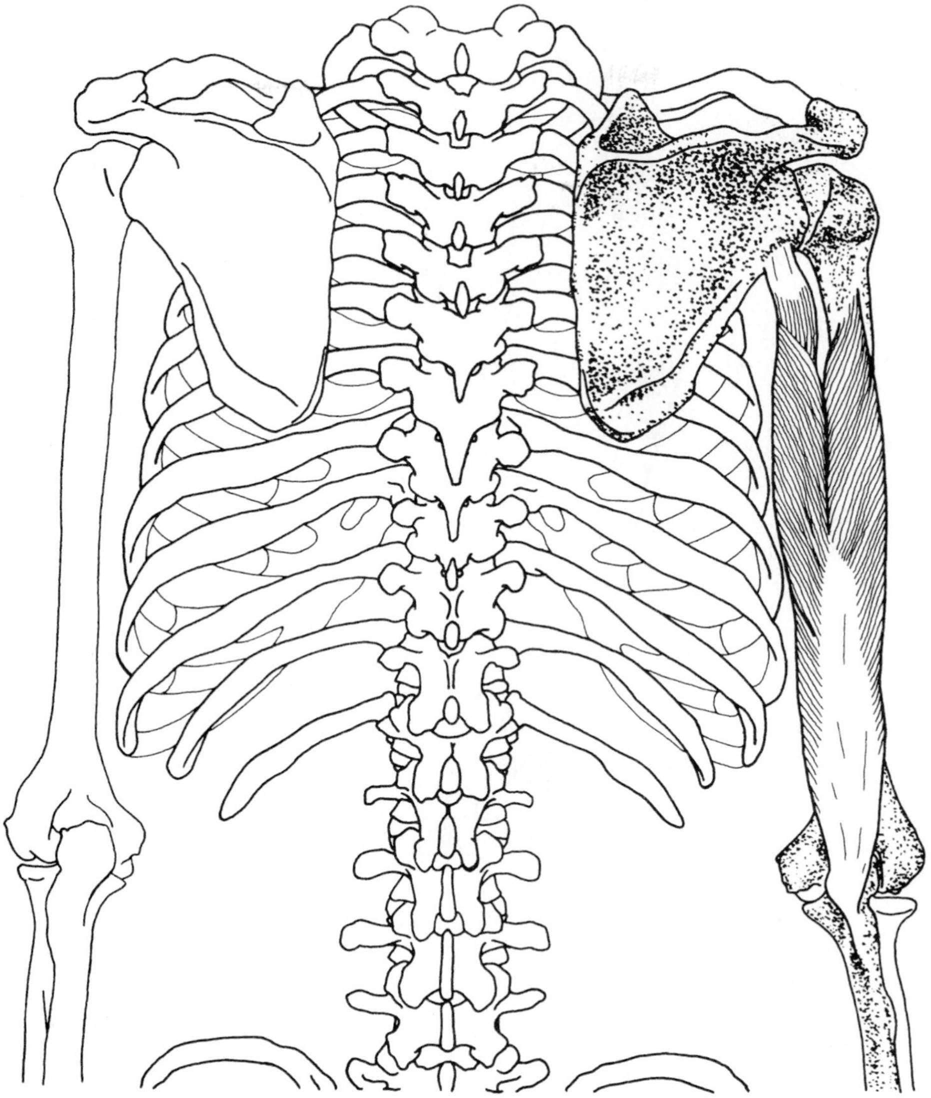

Posterior view

Origin

Long head—infraglenoid tubercle of the scapula

Lateral head—upper half of the posterior surface of the shaft of the humerus

Medial head—posterior surface of the lower half of the shaft of the humerus

Insertion

Posterior part of olecranon process of the ulna

Action

Extends forearm, long head aids in adduction if arm is abducted

Nerve

Radial nerve (C7, C8)

Note: The radial nerve comes from the axilla (armpit) and passes along the humerus between the medial and lateral heads. Because of its position, it is the most commonly injured peripheral nerve.

ANCONEUS

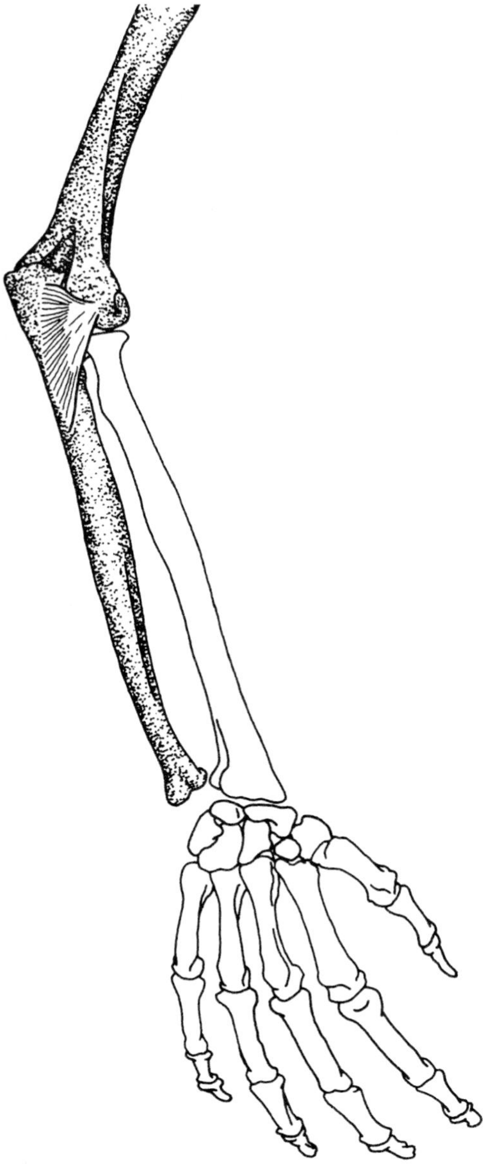

Posterior view of arm

Origin Posterior part of lateral epicondyle of the humerus

Insertion Lateral surface of the olecranon process and posterior surface of ulna

Action Extends forearm (assists triceps)

Nerve Radial nerve (C7, C8)

POSTERIOR BACK, SHOULDER, AND ARM MUSCLES

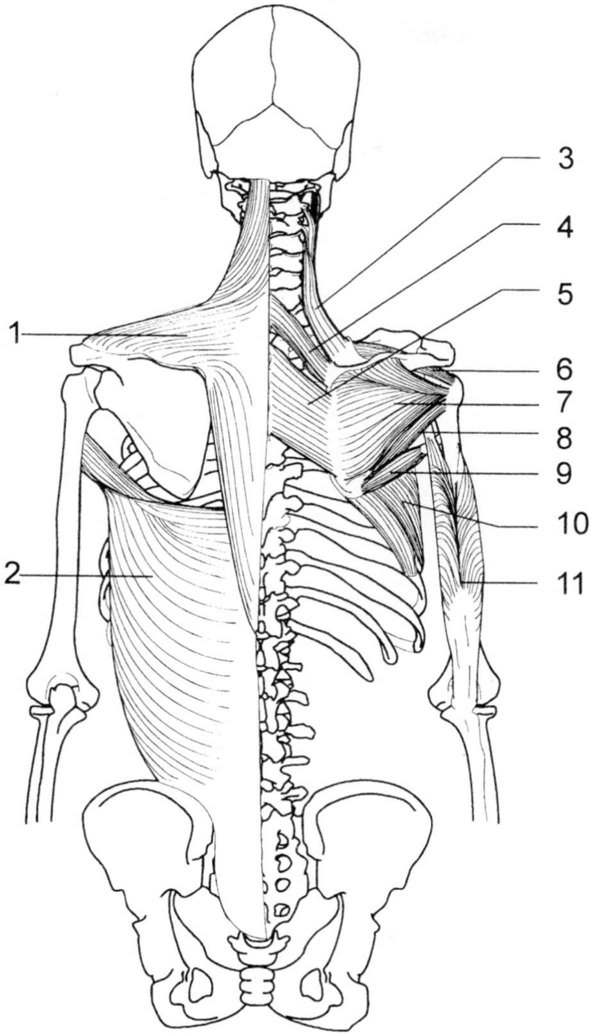

Trunk—dorsal view

Superficial layer

1. Trapezius
2. Latissimus dorsi

Deep layer

3. Levator scapulae
4. Rhomboideus minor
5. Rhomboideus major
6. Supraspinatus (rotator cuff)
7. Infraspinatus (rotator cuff)
8. Teres minor (rotator cuff)
9. Teres major
10. Serratus anterior

Posterior arm

11. Triceps brachii

CHAPTER SEVEN

MUSCLES OF THE FOREARM AND HAND

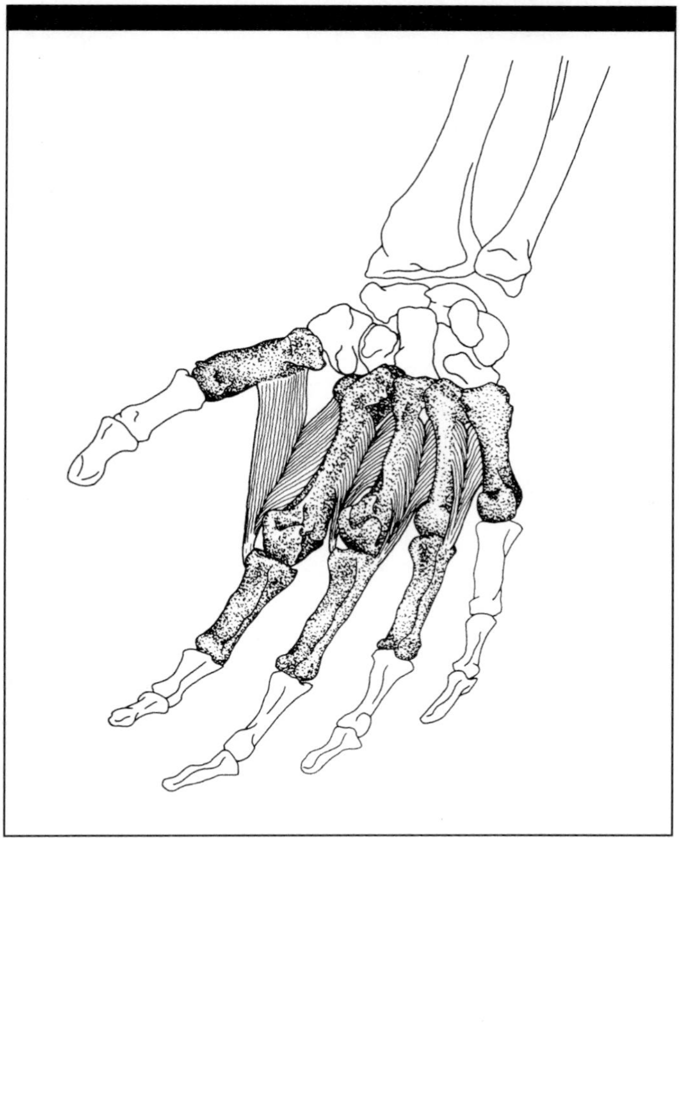

PRONATOR TERES

(Superficial group)

Origin Humeral head—medial supracondylar ridge and medial epicondyle of the humerus

Ulnar head—medial border of the coronoid process of the ulna

Insertion Middle of lateral surface of the radius (pronator tuberosity)

Action Pronates and flexes forearm

Nerve Median nerve (C6, C7)

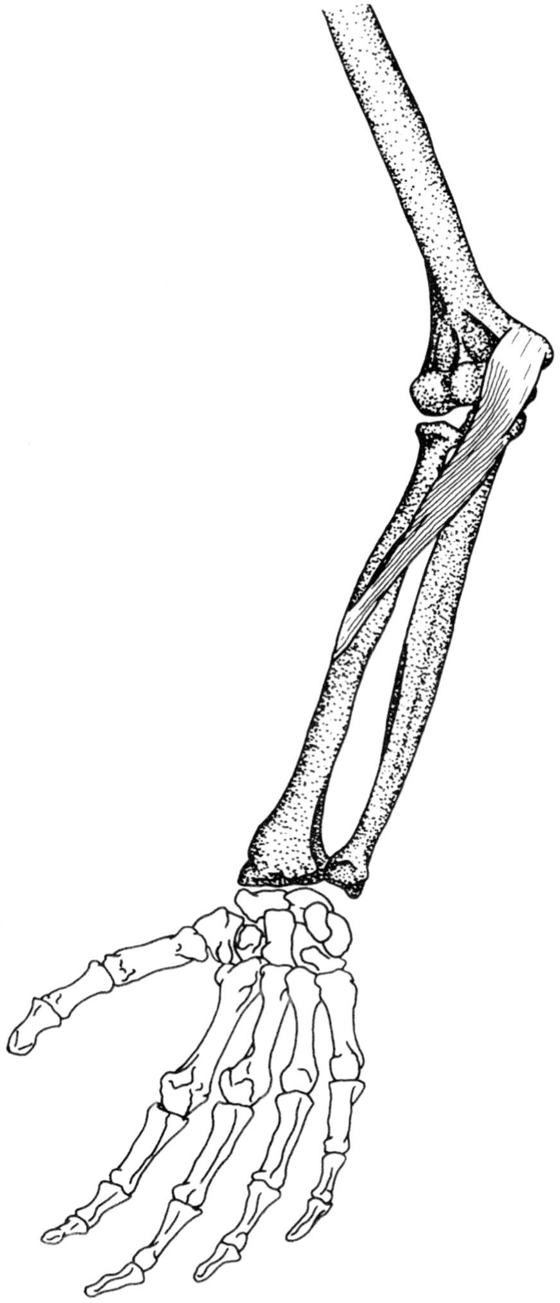

Forearm—anterior view

FLEXOR CARPI RADIALIS

(Superficial group)

Origin	Medial epicondyle of the humerus through the common tendon
Insertion	Front of the bases of the second and third metacarpal bones
Action	Flexes hand, synergist in abduction with extensor carpi radialis longus and brevis
Nerve	Median nerve (C6, C7)

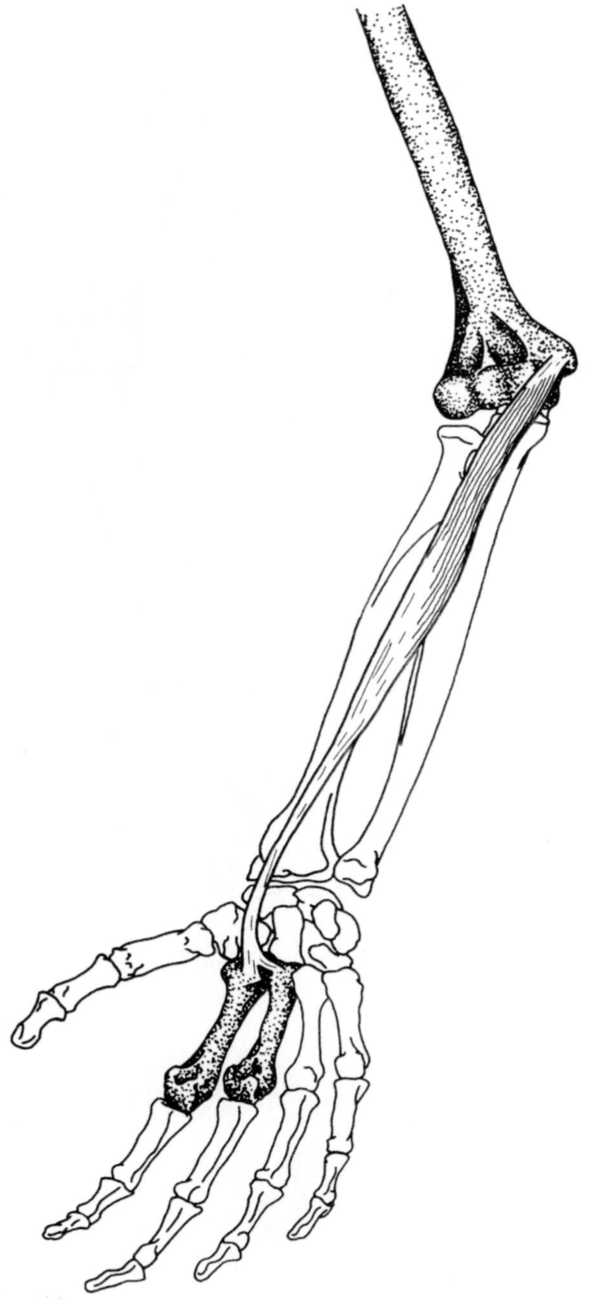

Forearm—anterior view

PALMARIS LONGUS

(Superficial group)

Origin	Medial epicondyle of the humerus through the common tendon
Insertion	Front (central part) of the flexor retinaculum and apex of the palmar aponeurosis
Action	Flexes the hand
Nerve	Median nerve (C6, C7)

Note: This muscle is absent in about 14% of limbs.

Reference: Agur, Amr: *Grant's Atlas of Anatomy,* 9th ed. Williams & Wilkins, Baltimore, 1991.

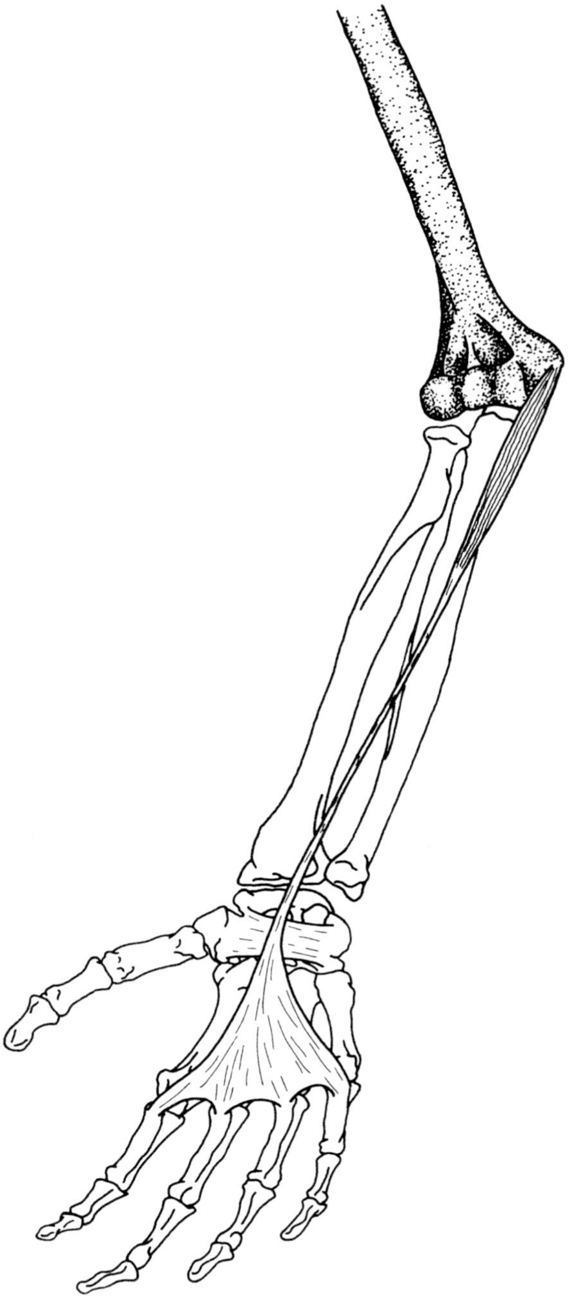

Forearm—anterior view

FLEXOR CARPI ULNARIS

(Superficial group)

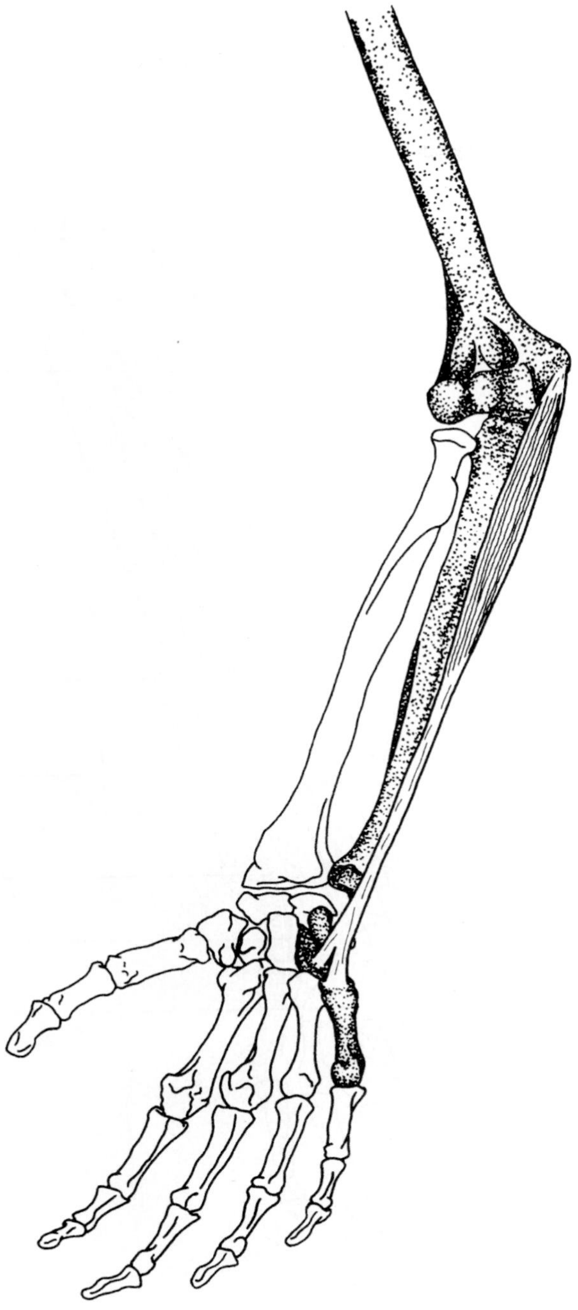

Forearm—anterior view

Origin Humeral head—medial epicondyle of the humerus through the common tendon

Ulnar head—medial margin of olecranon process of ulna, dorsal border of shaft of the ulna

Insertion Pisiform bone, hook of the hamate, and base of the fifth metacarpal bone

Action Flexes hand, synergist in adduction of hand with extensor carpi ulnaris

Nerve Ulnar nerve (C8, T1)

MUSCLES OF THE WRIST

1. Pronator teres
2. Flexor carpi radialis
3. Palmaris longus
4. Flexor carpi ulnaris
5. Flexor retinaculum
6. Palmar aponeurosis

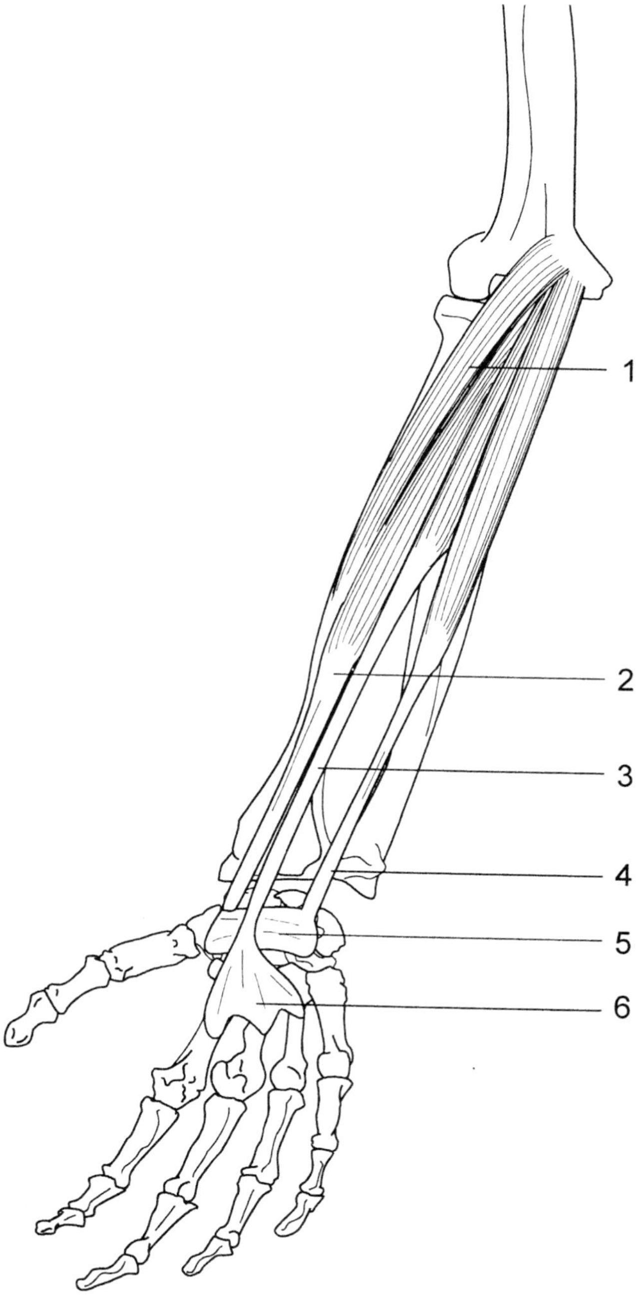

Forearm—anterior view

FLEXOR DIGITORUM SUPERFICIALIS

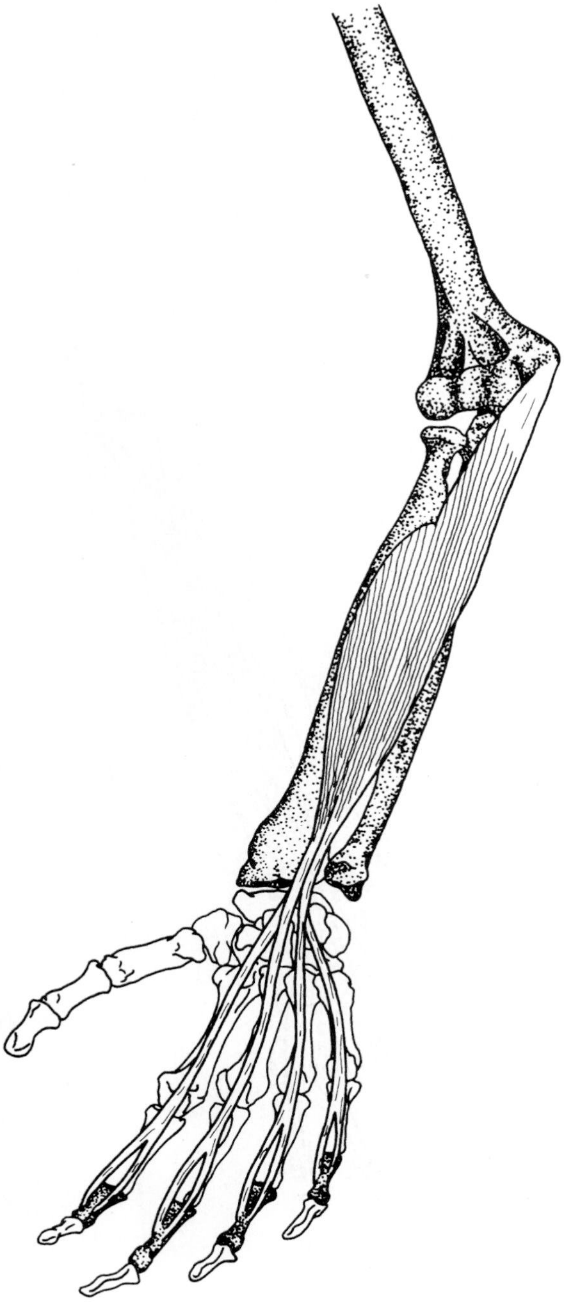

Forearm—anterior view

Origin	Humeroulnar head—medial epicondyle of the humerus through common tendon,* medial margin of the coronoid process of ulna Radial head—anterior surface of shaft of radius
Insertion	Four tendons divide into two slips each, slips insert into the sides (margins of the anterior surfaces) of the middle phalanges of four fingers
Action	Flexes the middle phalanges of the fingers
Nerve	Median nerve (C7, C8, T1)
Relationships	Deep to superficial flexors

*See superficial flexors.

FLEXOR DIGITORUM PROFUNDUS

Origin Upper three-fourths of anterior and medial surfaces of shaft of ulna and medial side of the coronoid process, interosseous membrane

Insertion Front of base of distal phalanges of fingers

Action Flexes distal phalanges

Nerve Ulnar nerve supplies the medial half of the muscle (going to the little and ring fingers) (C8, T1)

Anterior interosseous branch of median nerve supplies lateral half (going to index and middle fingers) (C8, T1)

Relationships Deep to flexor digitorum superficialis

Note: Both flexor digitorum muscles and the median nerve pass under the flexor retinaculum (page 10) in the wrist. When irritated, the synovial sheaths of these muscles can compress the median nerve, causing the sensory and motor deficits known as carpal tunnel syndrome.

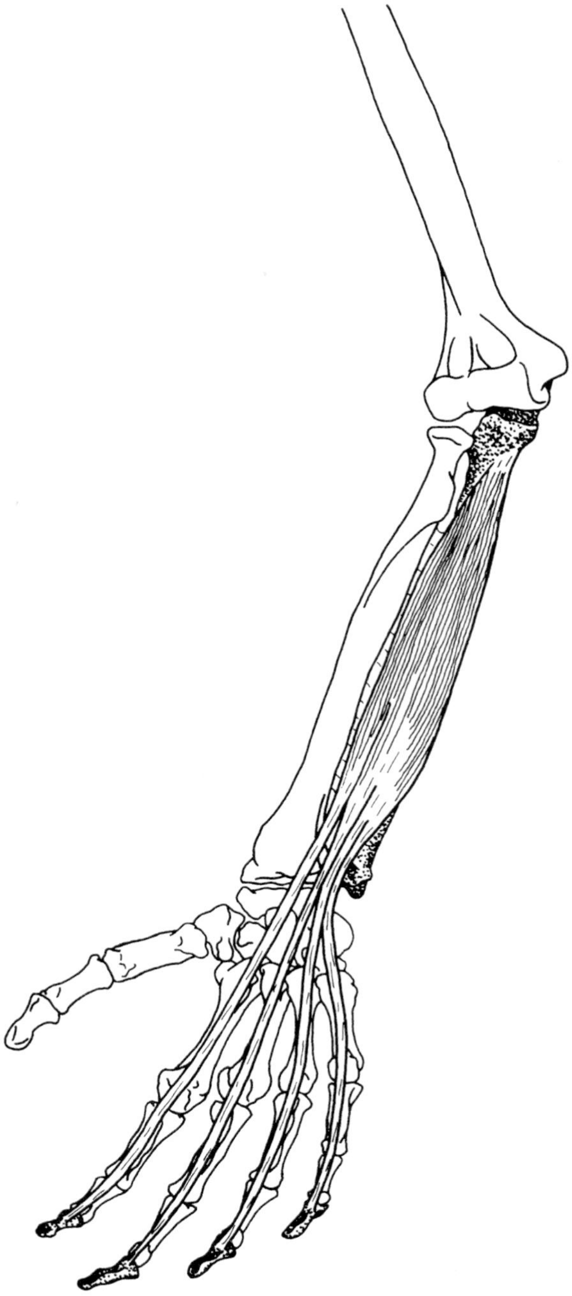

Forearm—anterior view

FLEXOR POLLICIS LONGUS

Origin Middle of anterior surface of shaft of radius, interosseous membrane, medial epicondyle of humerus, and often coronoid process of ulna

Insertion Palmar aspect of base of the distal phalanx of thumb

Action Flexes the thumb

Nerve Anterior interosseous branch of median nerve (C8, T1)

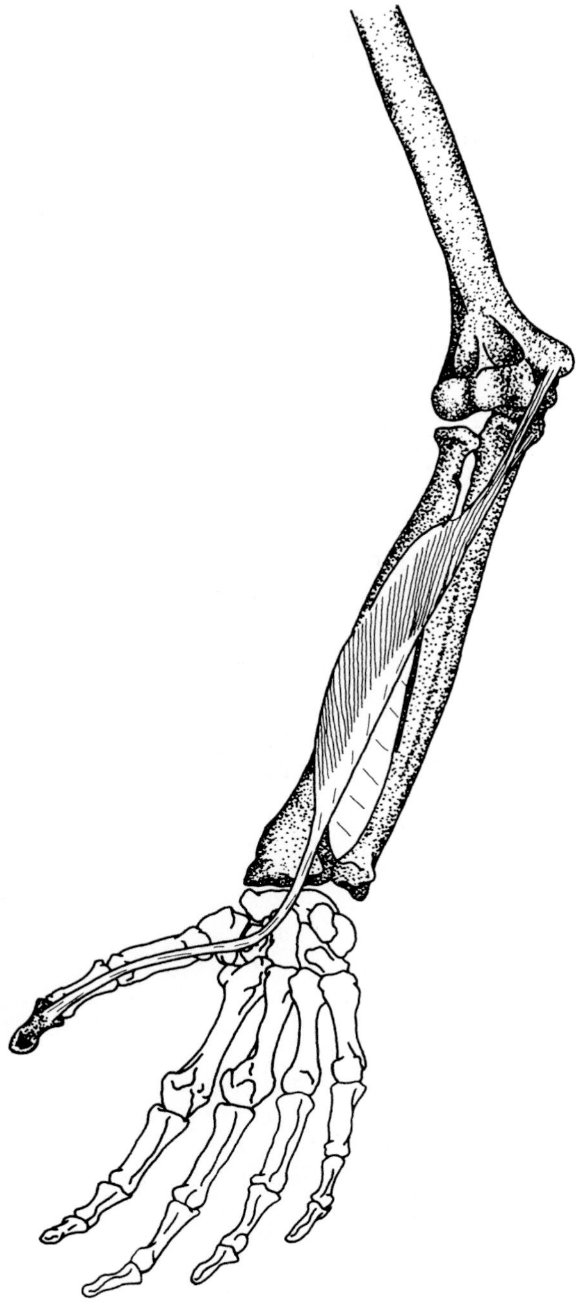

Forearm—anterior view

FLEXORS OF THE FINGERS

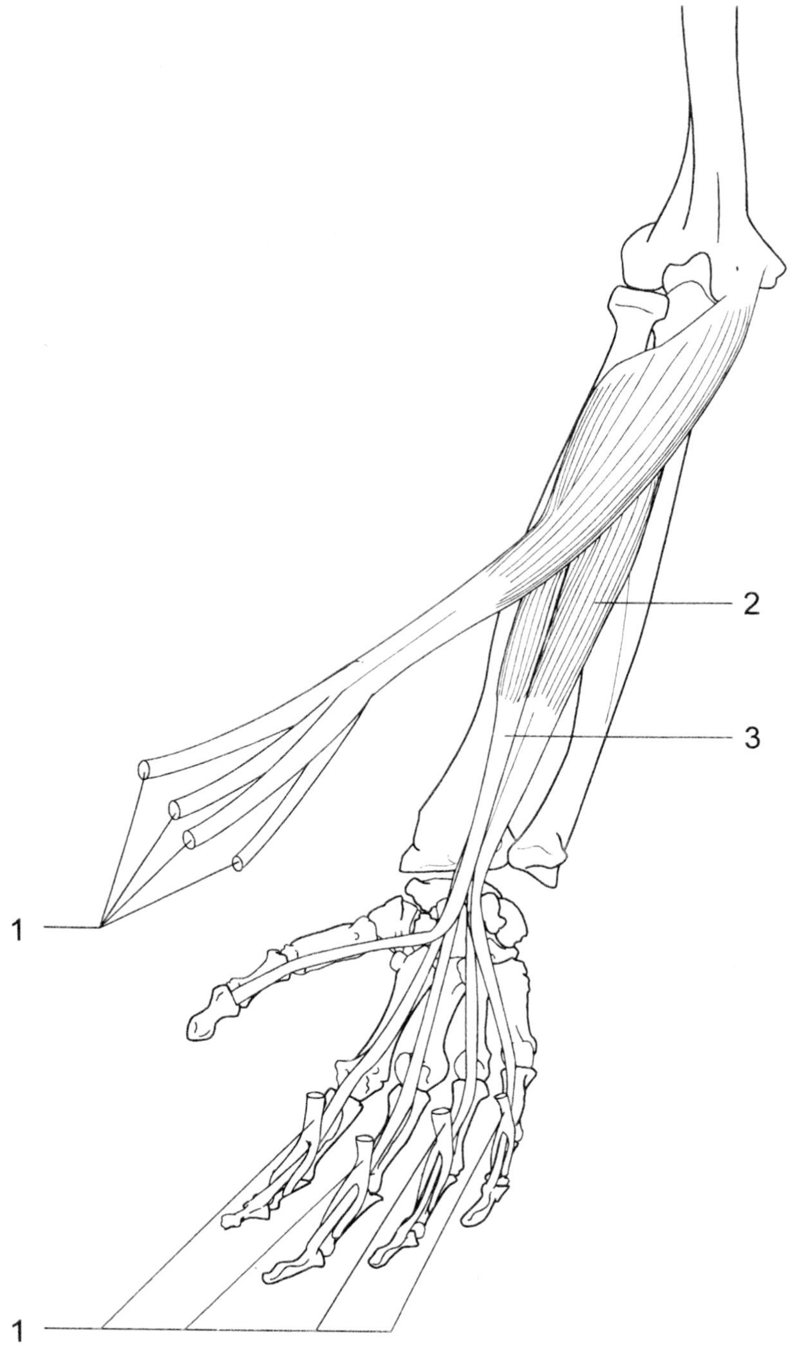

Forearm—anterior view

1. Flexor digitorum superficialis (cut)
2. Flexor digitorum profundus
3. Flexor pollicis longus

Note: The tendons of flexor digitorum superficialis split and attach to the middle phalanx. The tendons of flexor digitorum profundus pass through this split and continue to the distal phalanx.

PRONATOR QUADRATUS

Origin	Anterior surface of distal part of shaft of ulna
Insertion	Lower portion of anterior surface of shaft of radius, distal part of lateral border of radius
Action	Pronates forearm and hand
Nerve	Anterior interosseous branch of median nerve (C8, T1)
Relationships	Deepest forearm muscle

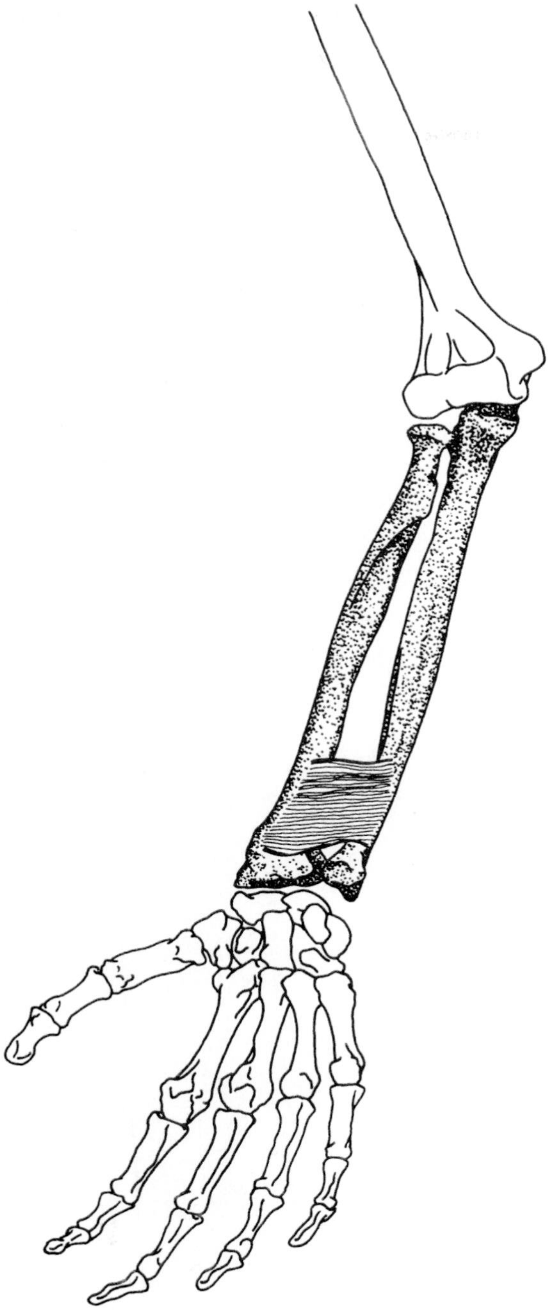

Forearm—anterior view

BRACHIORADIALIS

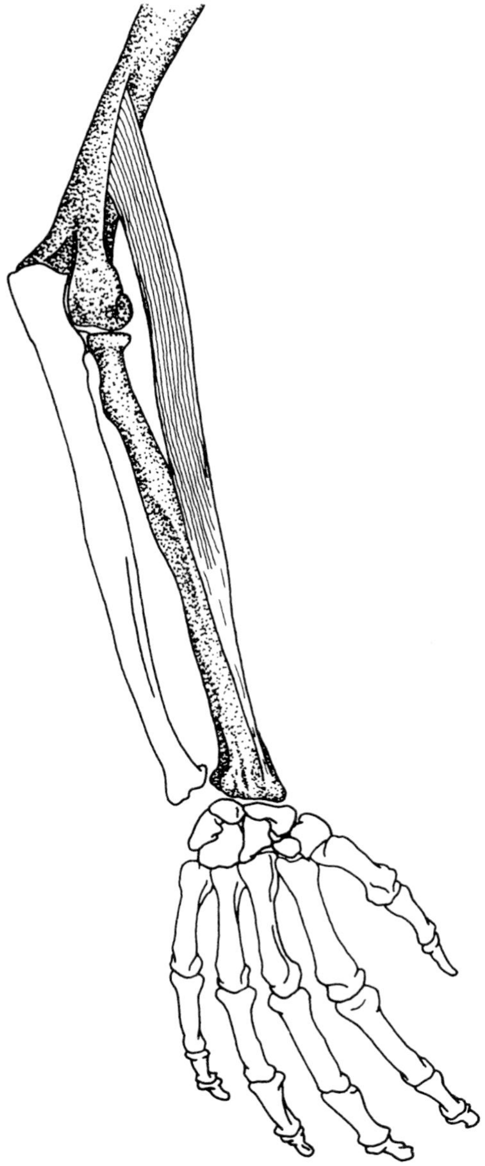

Forearm—dorsal view

Origin Upper two-thirds of lateral supracondylar ridge of humerus

Insertion Base of styloid process and lateral surface of radius

Action Flexes forearm

Nerve Radial nerve (C5, C6)

EXTENSOR CARPI RADIALIS LONGUS

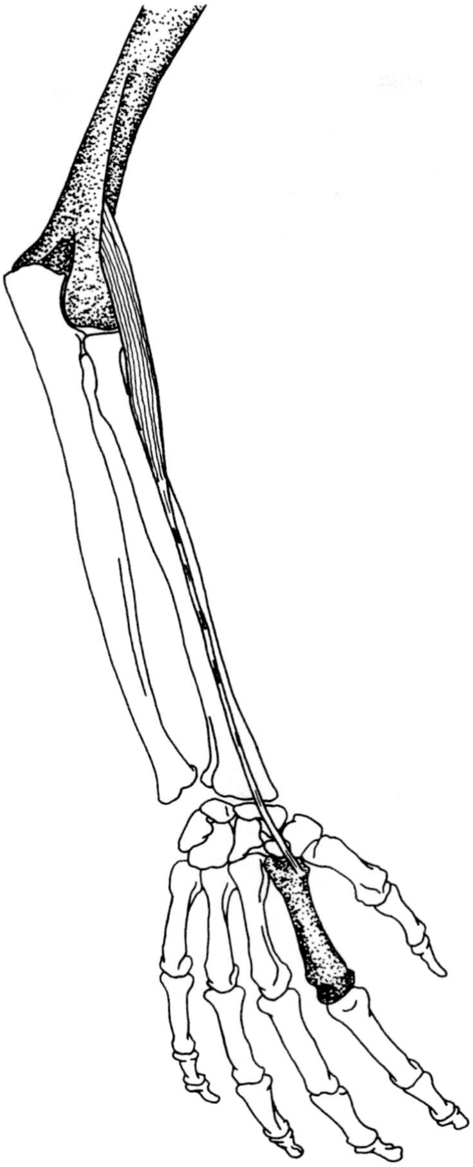

Forearm—dorsal view

Origin Lower third of lateral supracondylar ridge of humerus

Insertion Dorsal surface of the base of the second metacarpal bone

Action Extends hand, synergist in abduction of hand with flexor carpi radialis

Nerve Radial nerve (C6, C7)

EXTENSOR CARPI RADIALIS BREVIS

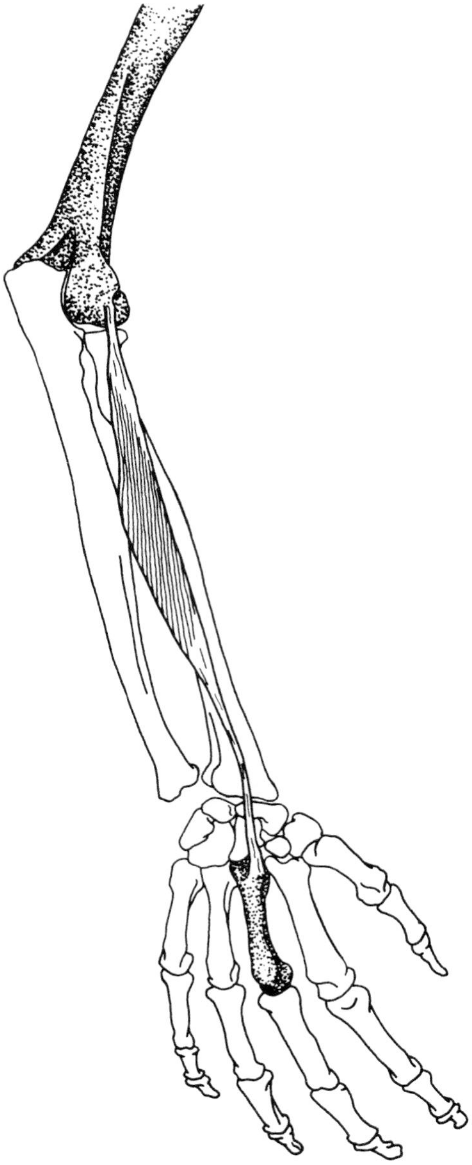

Forearm—dorsal view

Origin Lateral epicondyle of humerus

Insertion Dorsal surface of third metacarpal bone

Action Extends hand, synergist in abduction of hand with flexor carpi radialis

Nerve Radial nerve (C6, C7)

EXTENSOR DIGITORUM COMMUNIS

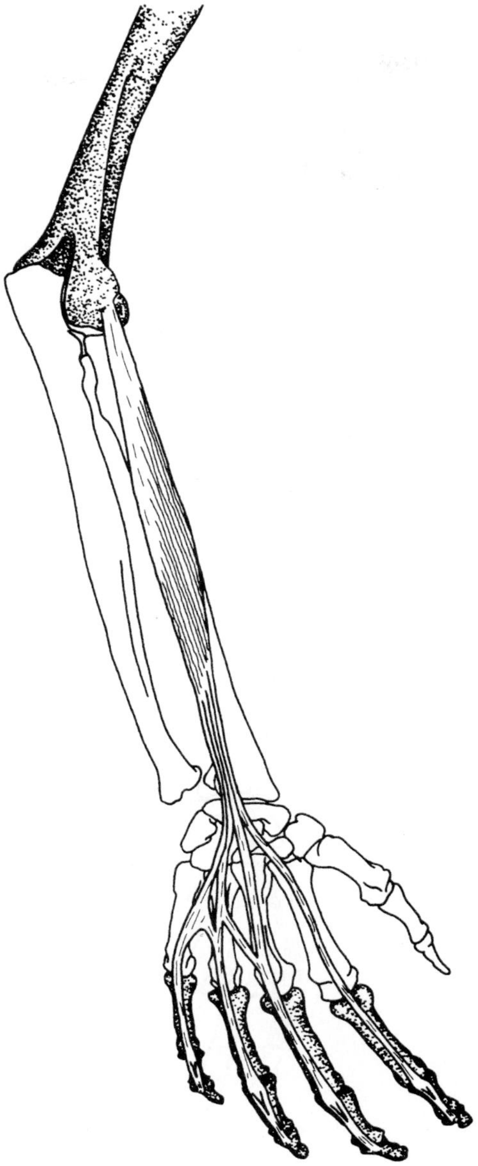

Forearm and hand—dorsal view

Origin Common tendon attached to lateral epicondyle of humerus

Insertion Lateral and dorsal surfaces of all the phalanges of the four fingers

Action Extends the fingers and wrist

Nerve Deep branch of radial nerve (C6–C8)

Relationships Tends to hyperextend the metacarpophalangeal joint causing "claw hand"; its action is balanced by the lumbricales and interossei

EXTENSOR DIGITI MINIMI

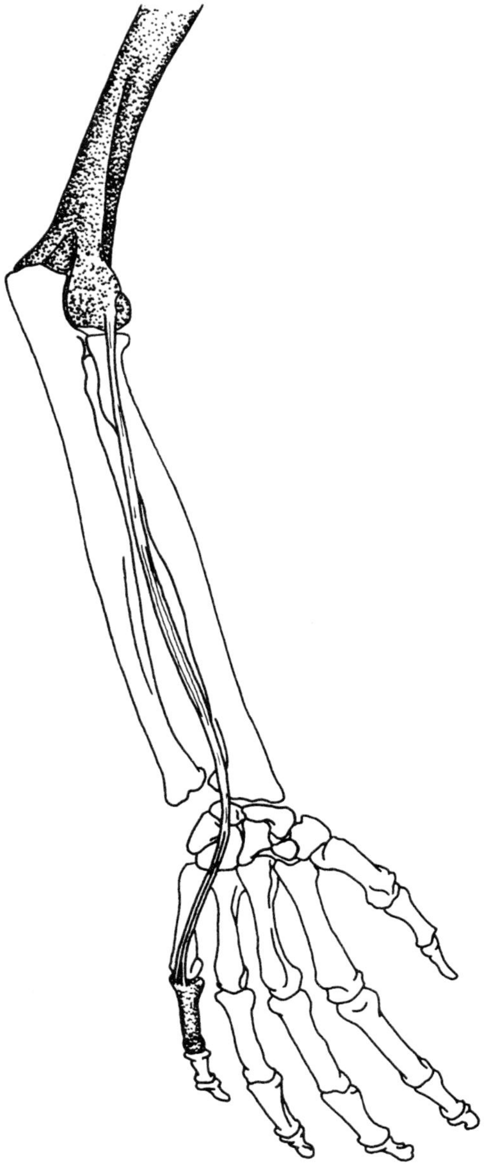

Forearm and hand—dorsal view

Origin Common tendon attached to lateral epicondyle of humerus

Insertion Dorsal surface of base of first phalanx of fifth finger

Action Extends fifth finger

Nerve Radial nerve (C6–C8)

EXTENSOR CARPI ULNARIS

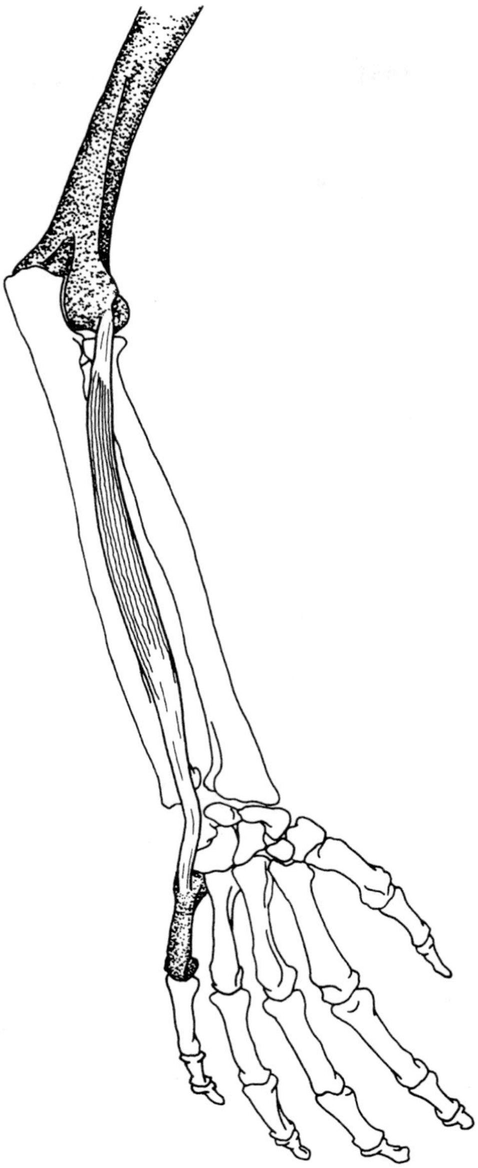

Forearm and hand—dorsal view

Origin Common tendon attached to lateral epicondyle of humerus

Insertion Dorsal surface of base of fifth metacarpal bone

Action Extends hand, synergist in adduction of hand with flexor carpi ulnaris

Nerve Radial nerve (C6–C8)

SUPINATOR

Origin Lateral epicondyle of humerus, lateral ligament (radial collateral) of elbow, annular ligament of superior radioulnar joint, supinator crest of ulna

Insertion Dorsal and lateral surfaces of upper third of radius

Action Supinates forearm

Nerve Radial nerve (C6)

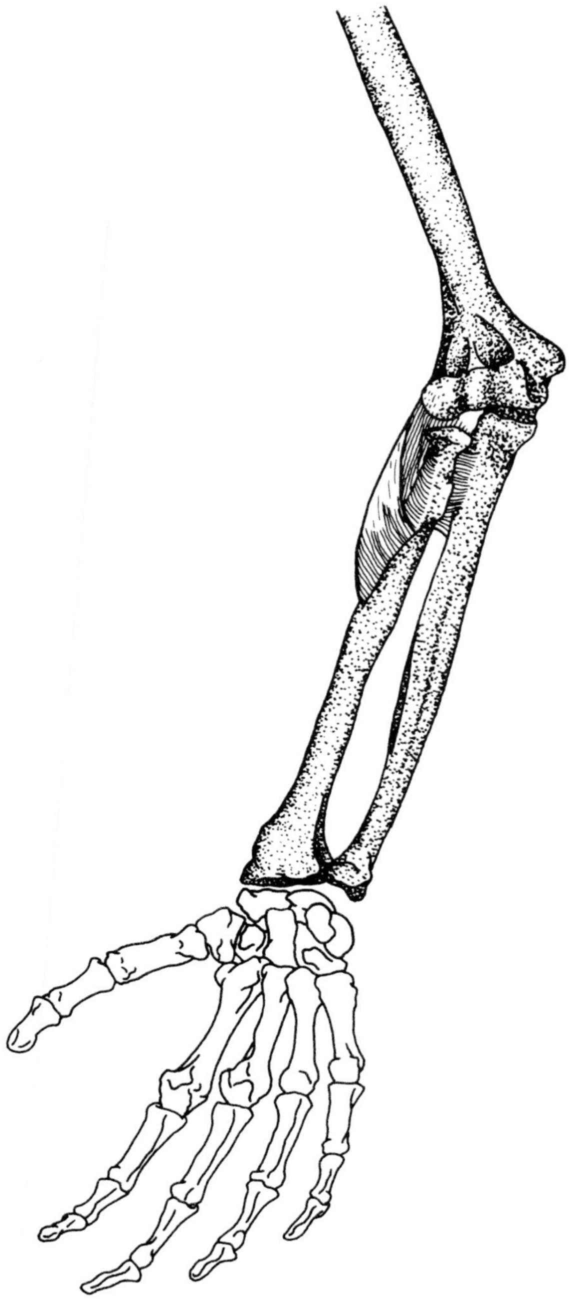

Forearm and hand—anterior view

ABDUCTOR POLLICIS LONGUS

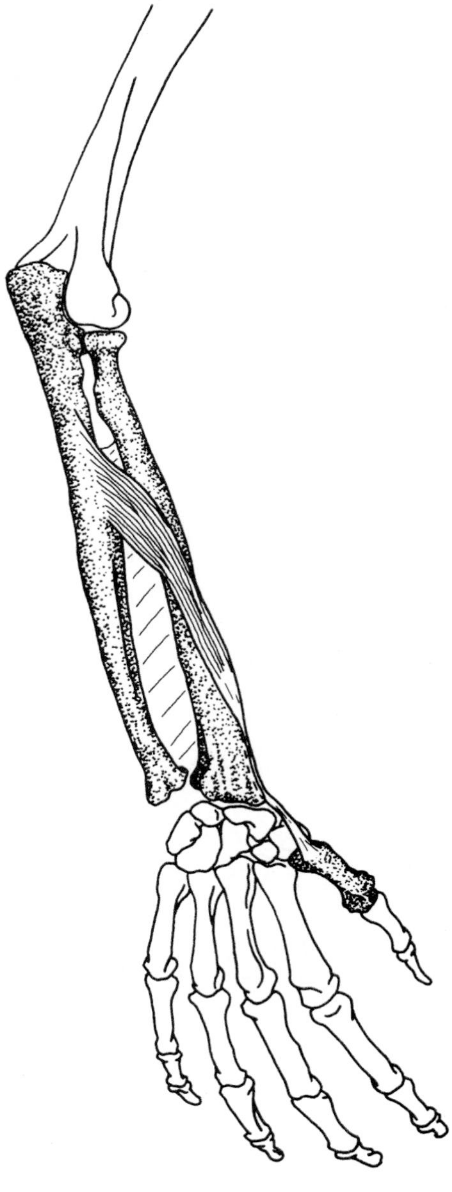

Forearm and hand—dorsal view

Origin Posterior (dorsal) surface of shaft of radius, ulna, interosseous membrane

Insertion Dorsal surface of base of first metacarpal bone

Action Abducts, laterally rotates, and extends thumb; abducts wrist

Nerve Radial nerve (C6, C7)

EXTENSOR POLLICIS BREVIS

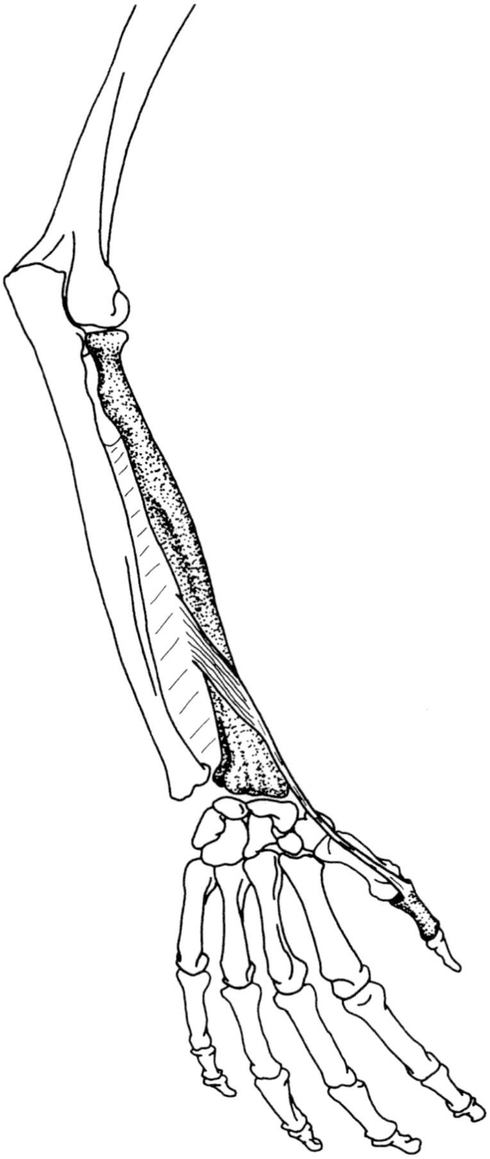

Forearm and hand—dorsal view

Origin Dorsal surface of radius, adjacent part of interosseous membrane

Insertion Base of proximal phalanx of thumb

Action Extends thumb, abducts hand

Nerve Radial nerve (C6, C7)

EXTENSOR POLLICIS LONGUS

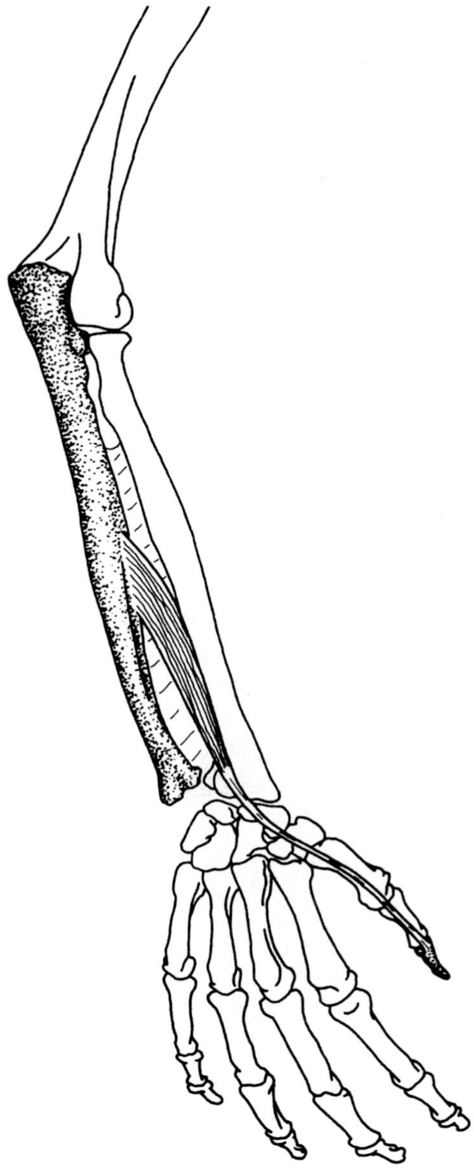

Forearm and hand—dorsal view

Origin Middle third of dorsal surface of ulna, interosseous membrane

Insertion Base of distal phalanx of thumb

Action Extends thumb

Nerve Radial nerve (C6–C8)

EXTENSORS OF THE THUMB

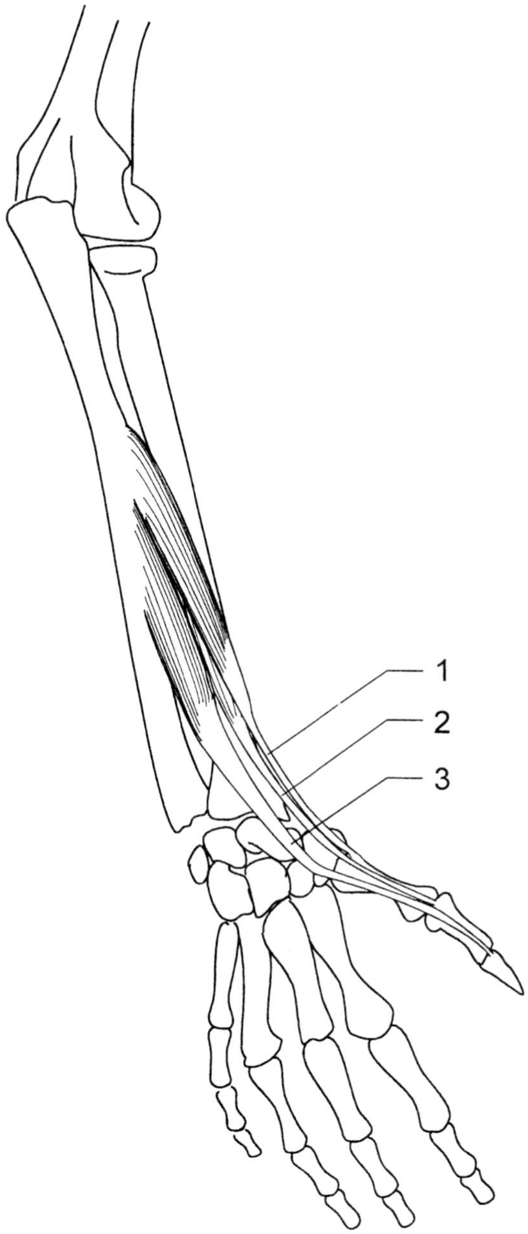

Forearm—posterior view

1. Abductor pollicis longus
2. Extensor pollicis brevis
3. Extensor pollicis longus

EXTENSOR INDICIS

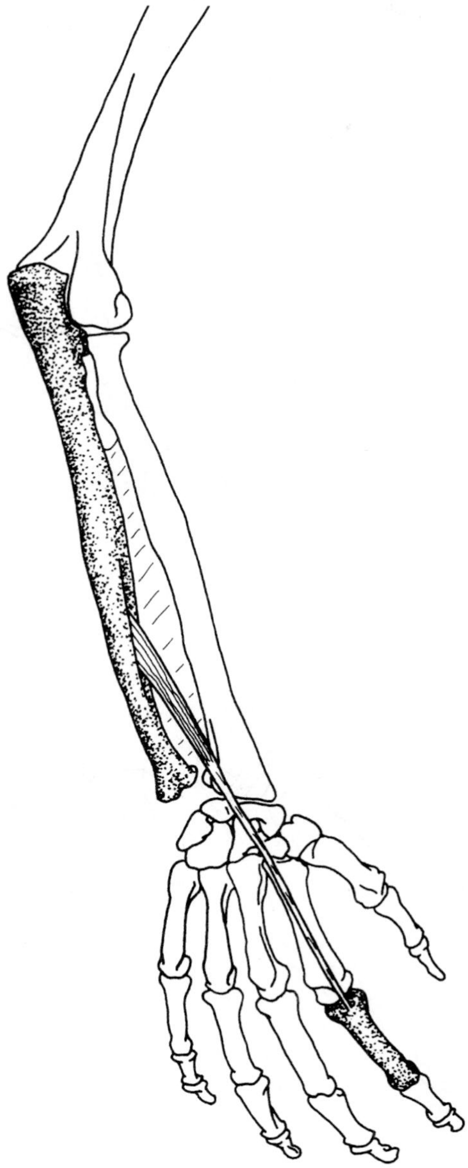

Forearm and hand—dorsal view

Origin Posterior surface of ulna and adjacent part of interosseous membrane

Insertion Extensor expansion on dorsal surface of proximal phalanx of index finger

Action Extends index finger

Nerve Radial nerve (C6–C8)

PALMARIS BREVIS

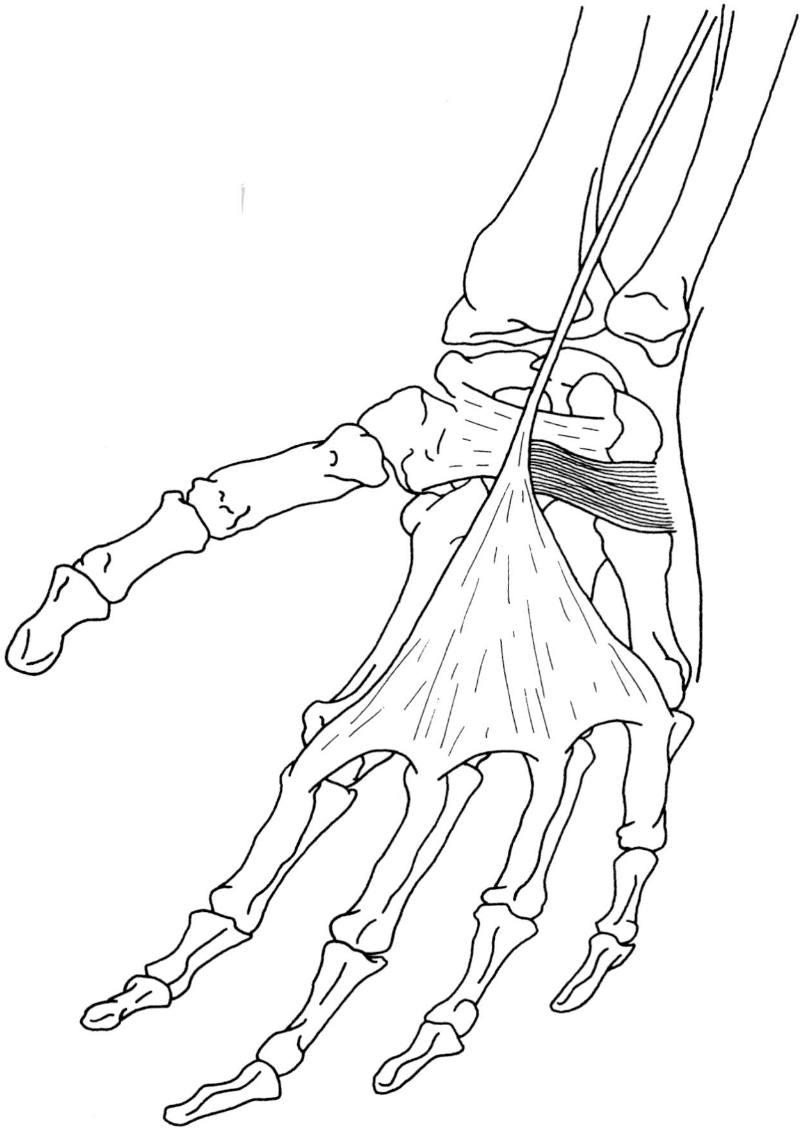

Hand—palmar view

Origin Flexor retinaculum, palmar aponeurosis

Insertion Skin of the palm

Action Corrugates skin of palm

Nerve Ulnar nerve (C8)

ABDUCTOR POLLICIS BREVIS

(Thenar eminence)

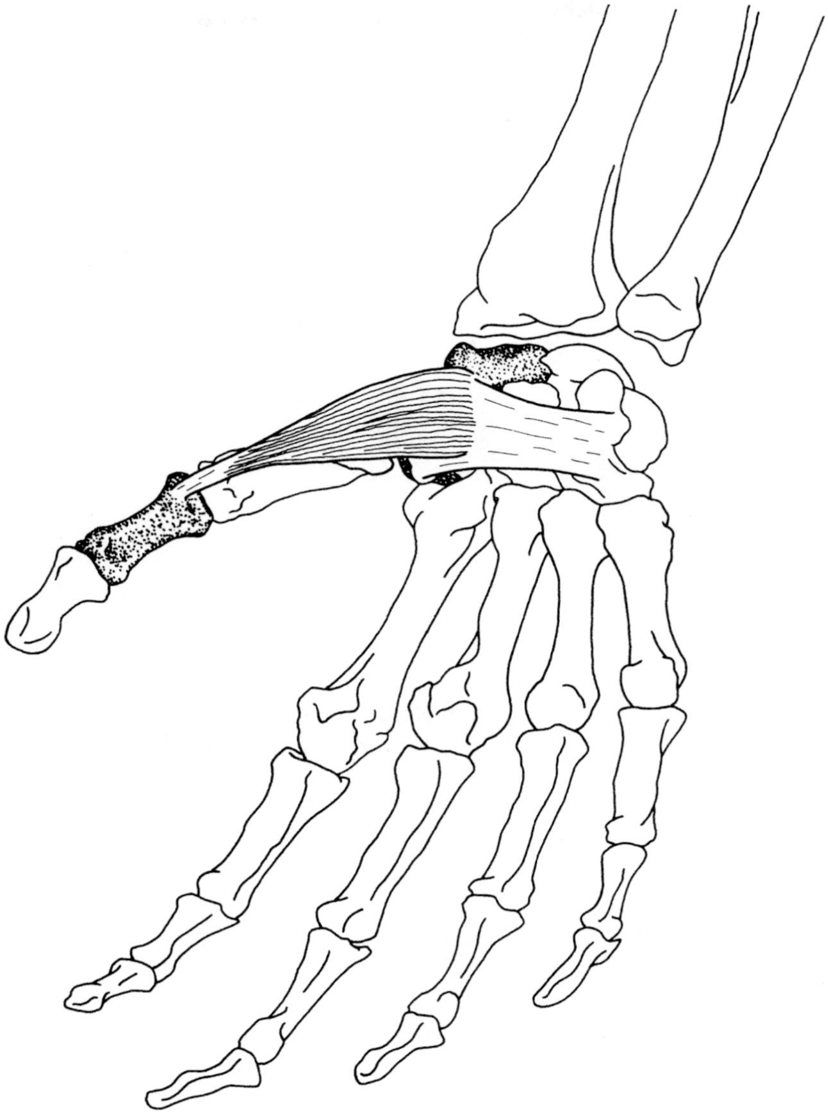

Hand—palmar view

Origin Tubercle of scaphoid, tubercle of trapezium, flexor retinaculum

Insertion Base of proximal phalanx of thumb

Action Abducts thumb and moves it anteriorly, acts together with other muscles of thenar eminence to oppose thumb to other fingers

Nerve Median (C6, C7)

Note: The abductor pollicis brevis, flexor pollicis brevis, and opponens pollicis form the thenar eminence at the base of the thumb.

FLEXOR POLLICIS BREVIS

(Thenar eminence)

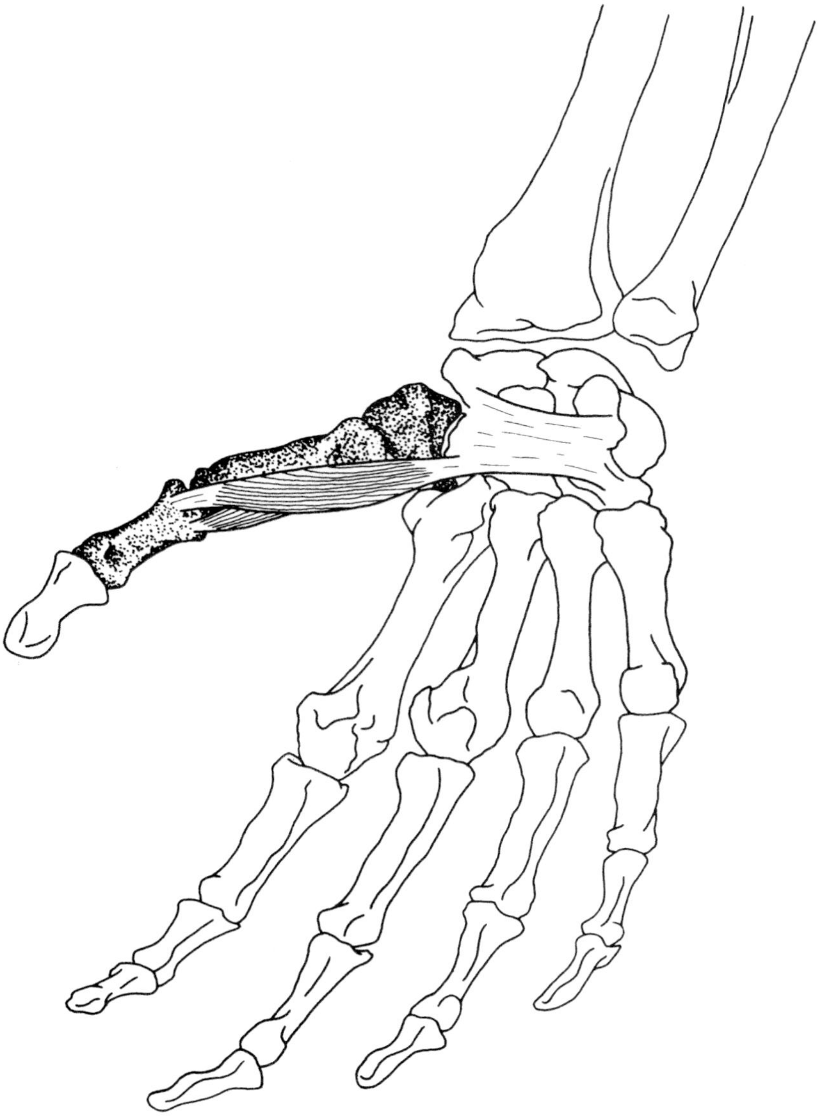

Hand—palmar view

Origin Flexor retinaculum and trapezium, and first metacarpal bone

Insertion Base of proximal phalanx of thumb

Action Flexes metacarpophalangeal joint of thumb, assists in abduction and rotation of thumb, acts together with other muscles of thenar eminence to oppose thumb to other fingers

Nerve Lateral portion—median nerve (C6, C7)

Medial portion—ulnar nerve (C8, T1)

OPPONENS POLLICIS

(Thenar eminence)

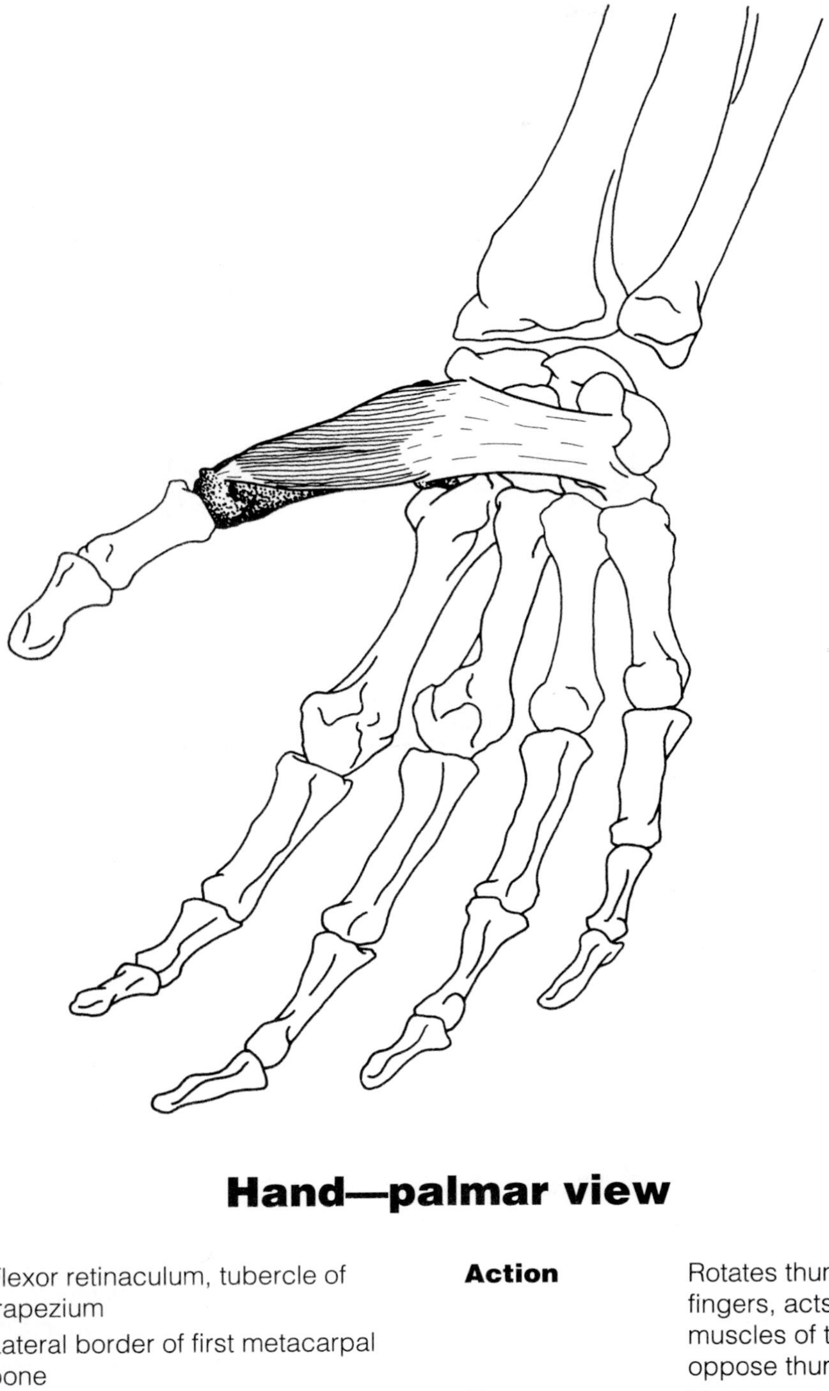

Hand—palmar view

Origin Flexor retinaculum, tubercle of trapezium

Insertion Lateral border of first metacarpal bone

Action Rotates thumb into opposition with fingers, acts together with other muscles of thenar eminence to oppose thumb to other fingers

Nerve Median nerve (C6, C7)

ADDUCTOR POLLICIS

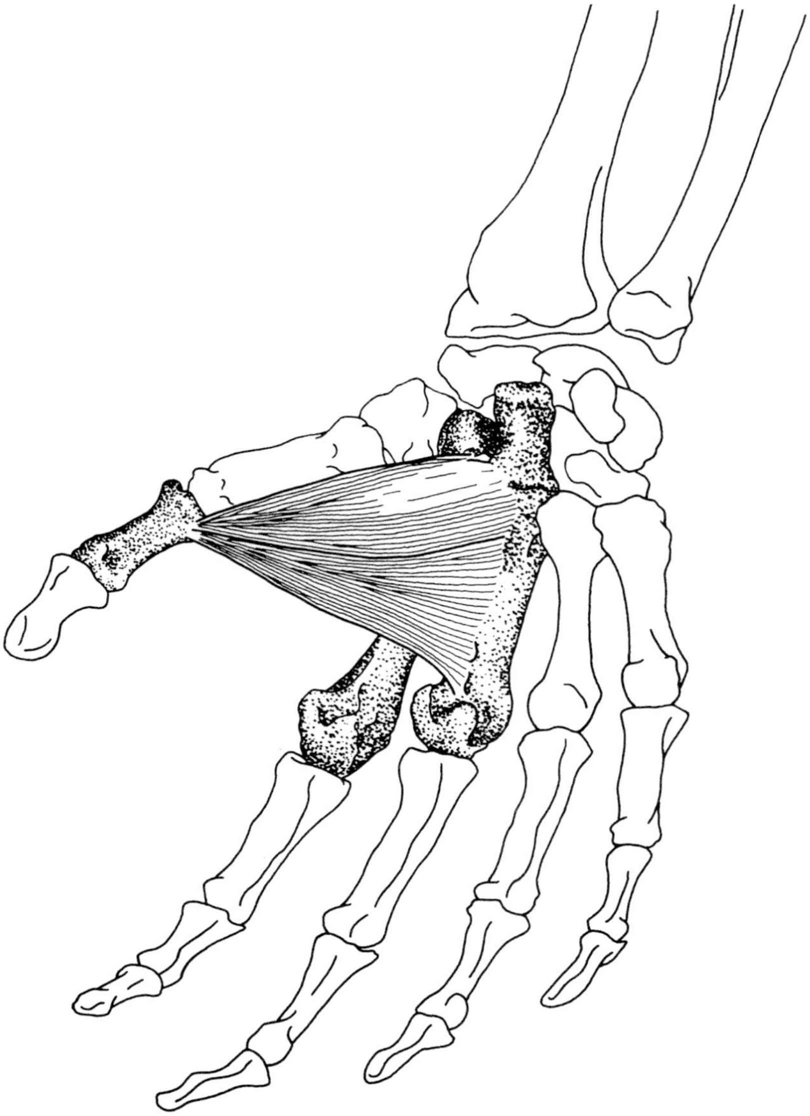

Hand—palmar view

Origin
Oblique head—anterior surfaces of second and third metacarpals, capitate, trapezoid
Transverse head—anterior surface of third metacarpal bone

Insertion
Medial side of base of proximal phalanx of the thumb

Action
Adducts thumb

Nerve
Ulnar nerve (C8, T1)

ABDUCTOR DIGITI MINIMI

(Hypothenar eminence)

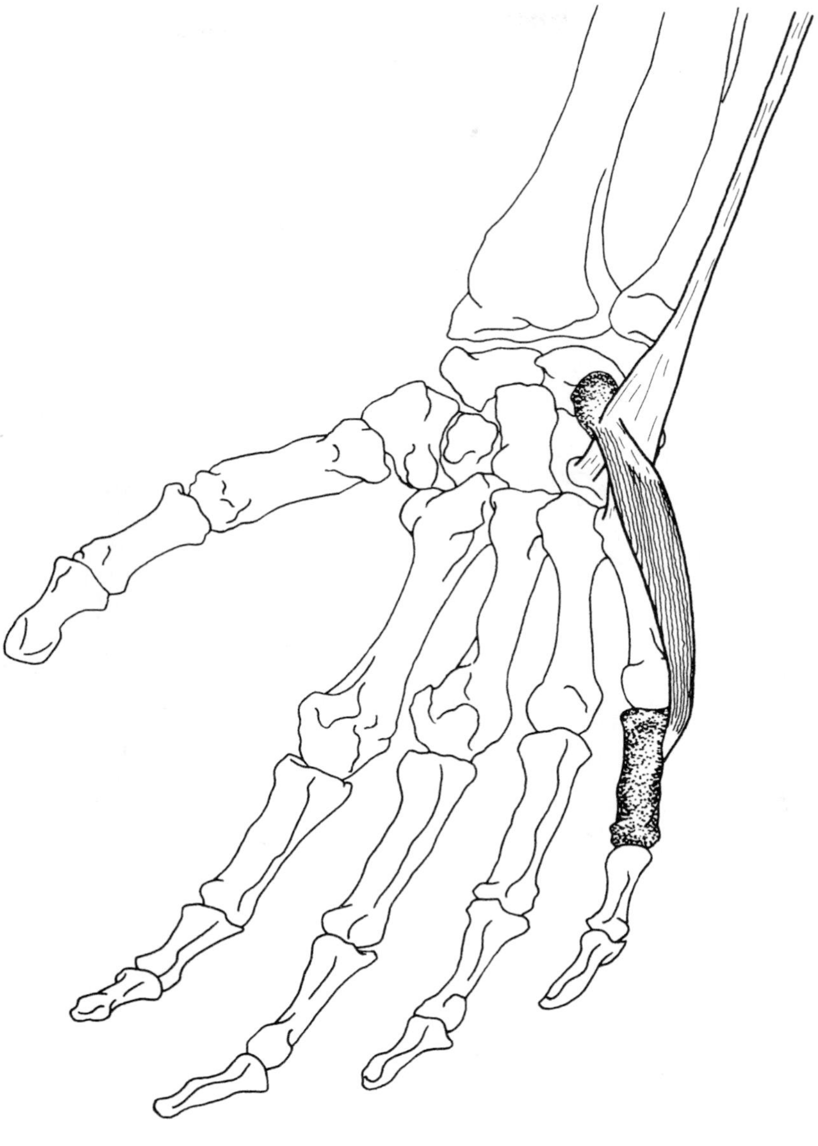

Hand—palmar view

Origin Pisiform bone, tendon of flexor carpi ulnaris

Insertion Medial side of base of proximal phalanx of fifth finger

Action Abducts fifth finger

Nerve Ulnar nerve (C8, T1)

Note: The hypothenar eminence is less prominent than the thenar eminence, and the fifth finger obviously cannot oppose the other digits.

FLEXOR DIGITI MINIMI BREVIS

(Hypothenar eminence)

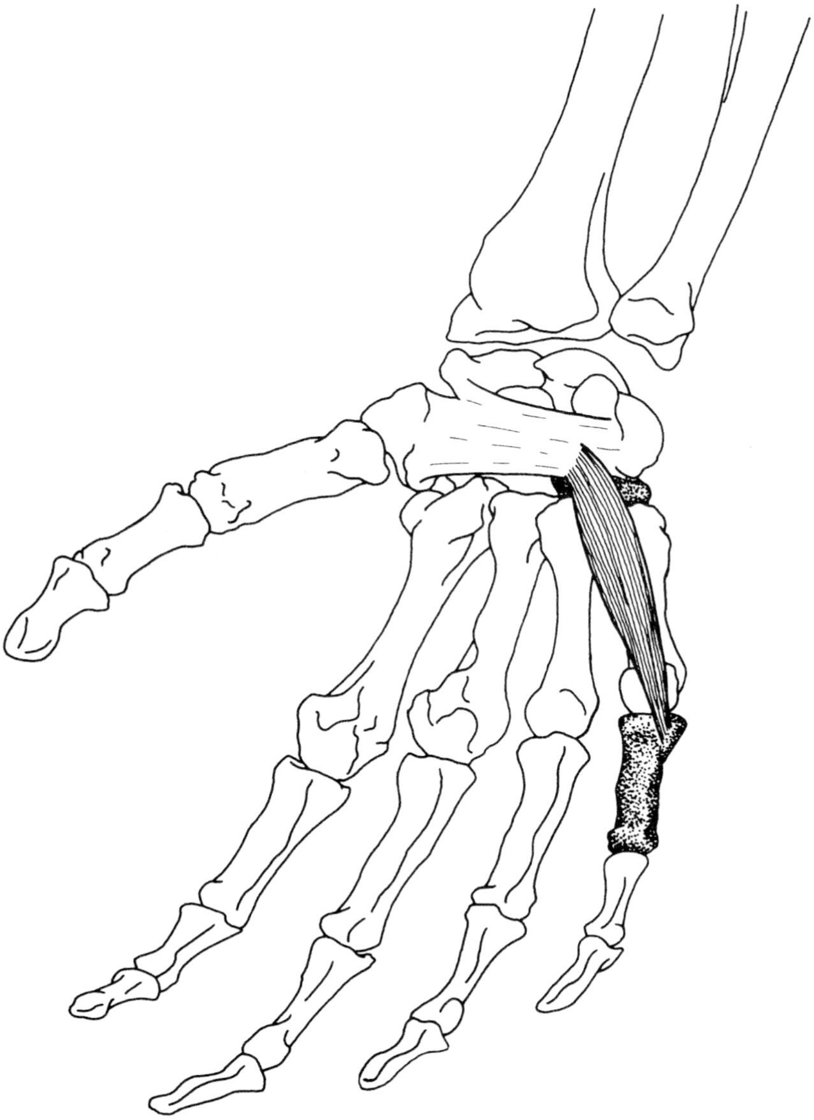

Hand—palmar view

Origin Anterior surface of flexor retinaculum, hook of hamate

Insertion Medial side of base of proximal phalanx of fifth finger

Action Flexes fifth finger at metacarpophalangeal joint

Nerve Ulnar nerve (C8, T1)

OPPONENS DIGITI MINIMI

(Hypothenar eminence)

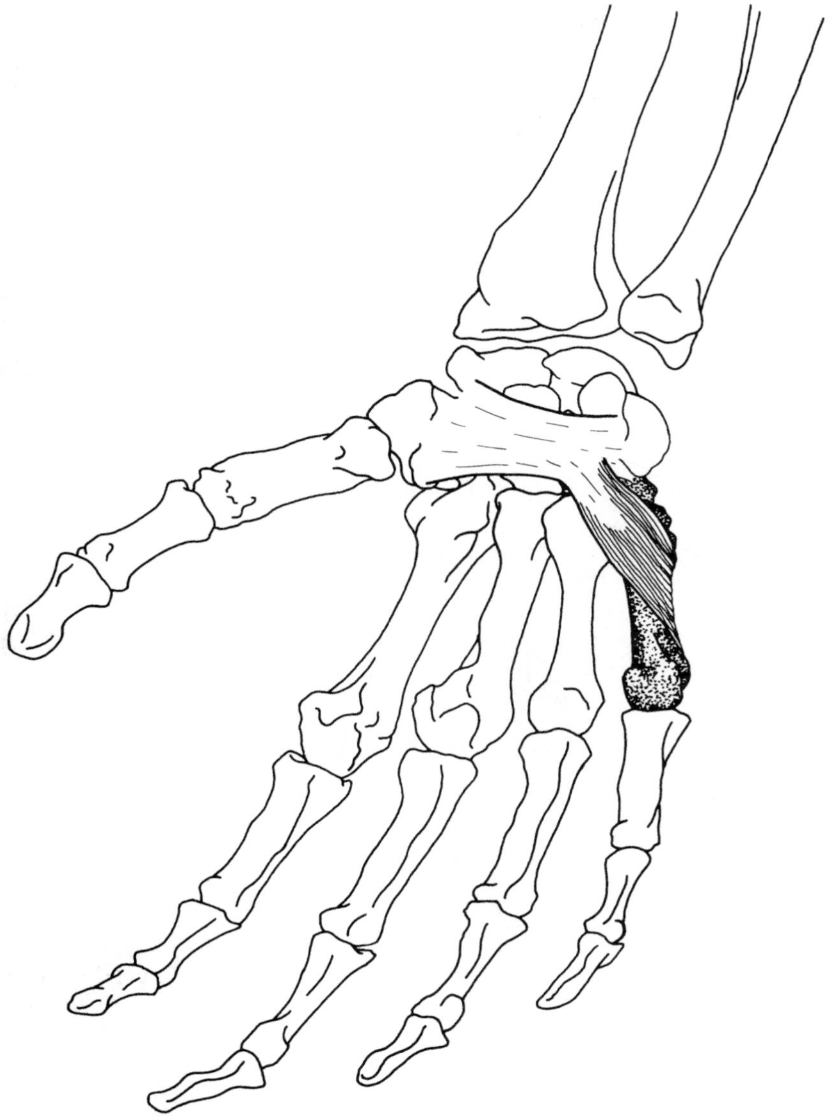

Hand—palmar view

Origin Anterior surface of flexor retinaculum, hook of hamate

Insertion Whole length of medial border of fifth metacarpal bone

Action Rotates fifth metacarpal bone, draws fifth metacarpal bone forward, assists flexor digiti minimi brevis in flexing carpometacarpal joint of fifth finger

Nerve Ulnar nerve (C8, T1)

LUMBRICALES*

(Four muscles)

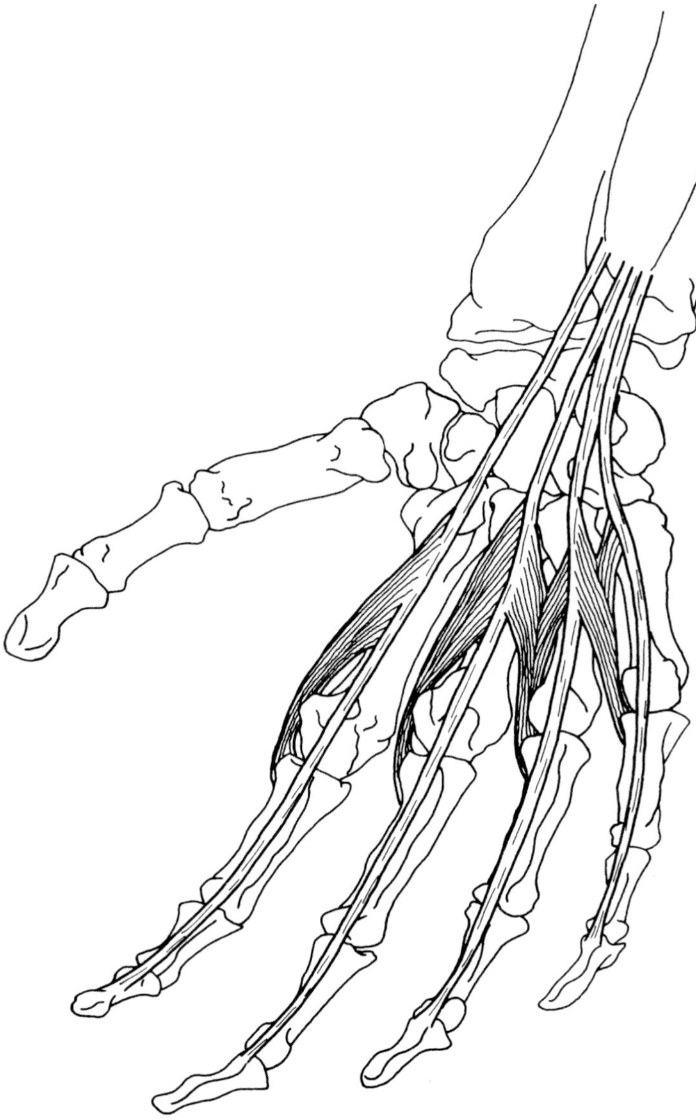

Hand—palmar view

Origin Tendons of flexor digitorum profundus in palm

Insertion Lateral side of corresponding tendon of extensor digitorum on fingers

Action Extend fingers at interphalangeal joints, weakly flex fingers at metacarpophalangeal joints

Nerve Lateral lumbricals (first and second)—median nerve (C6, C7)

Medial lumbricals (third and fourth)—ulnar nerve (C8)

Relationships Assist extensor digitorum communis in extending fingers without hyperextension at the metacarpophalangeal joints

*Associated with the tendons of flexor digitorum profundus.

PALMAR INTEROSSEI

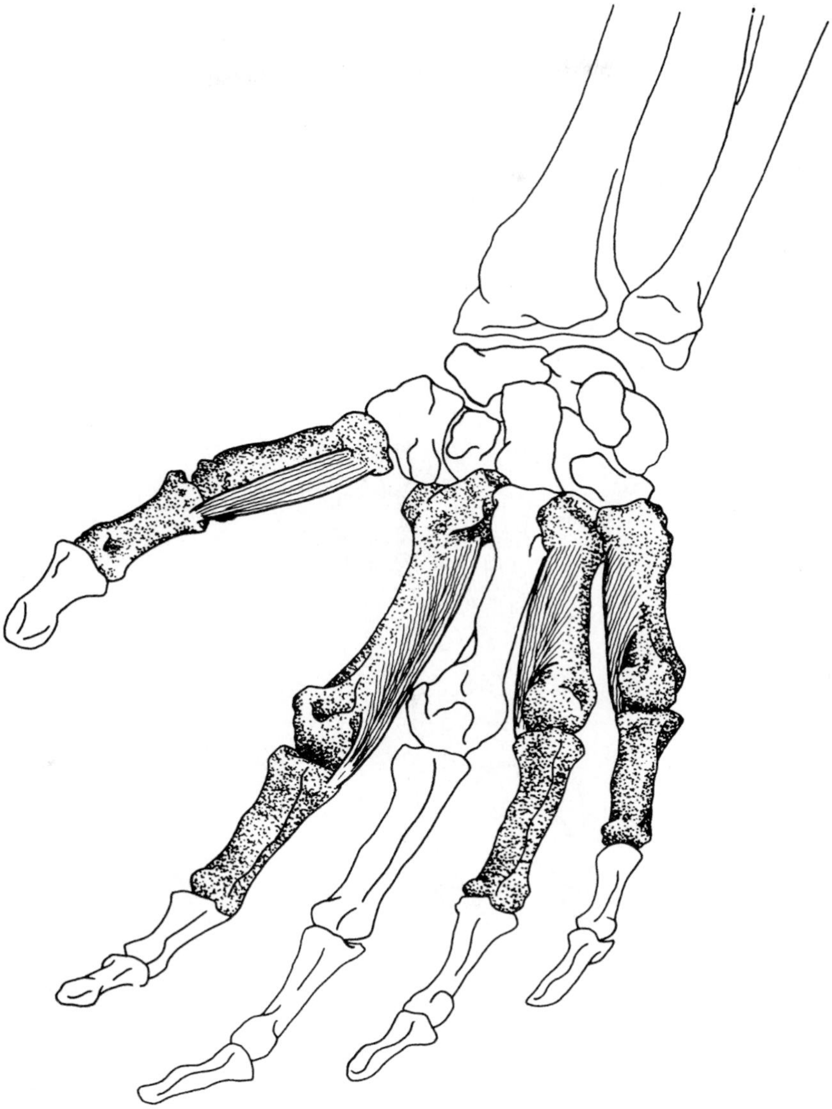

Hand—palmar view

Origin First—medial side of base of first metacarpal bone
Second, third, and fourth—anterior surfaces of second, fourth, and fifth metacarpal bones

Insertion First—medial side of base of proximal phalanx of thumb
Second—medial side of base of proximal phalanx of index finger
Third and fourth—lateral side of proximal phalanges of ring finger and fifth finger

Action Adduct fingers toward center of third finger at metacarpophalangeal joints, assist in flexion of fingers at metacarpophalangeal joints

Nerve Ulnar nerve (C8, T1)

Note: The palmar interosseus of the thumb, called the palmar interosseus of Henle, is usually absent.

DORSAL INTEROSSEI

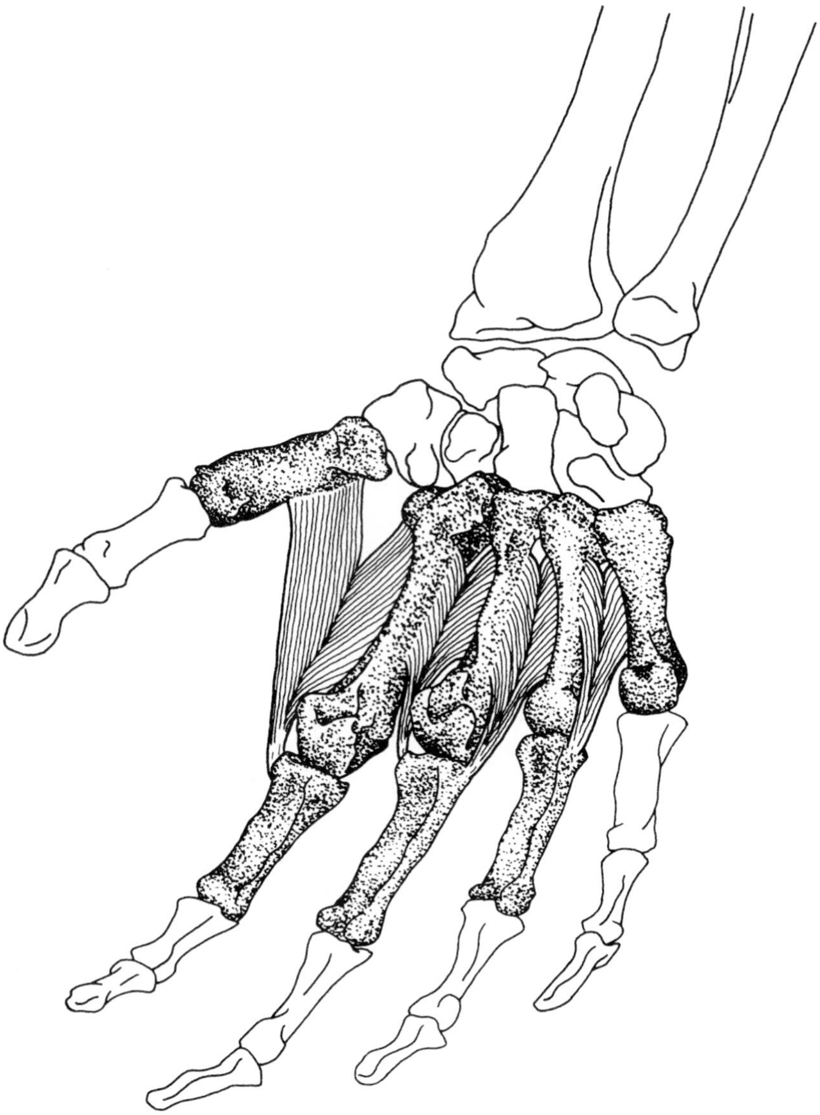

Hand—palmar view

Origin By two heads from adjacent sides of first and second, second and third, third and fourth, and fourth and fifth metacarpal bones

Insertion First—lateral side of base of proximal phalanx of index finger

Second—lateral side of base of proximal phalanx of middle finger

Third—medial side of base of proximal phalanx of middle finger

Fourth—medial side of base of proximal phalanx of ring finger

Action Abduct fingers away from center of third finger at metacarpophalangeal joints, assist in flexion of fingers at metacarpophalangeal joints

Nerve Ulnar nerve (C8, T1)

CHAPTER EIGHT

MUSCLES OF THE HIP AND THIGH

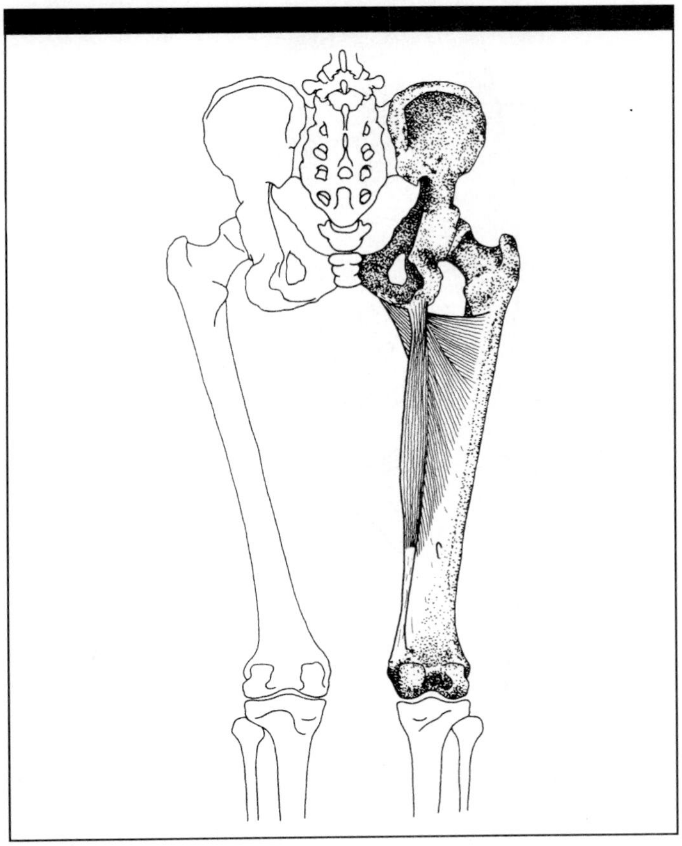

PSOAS MAJOR

(Part of iliopsoas)

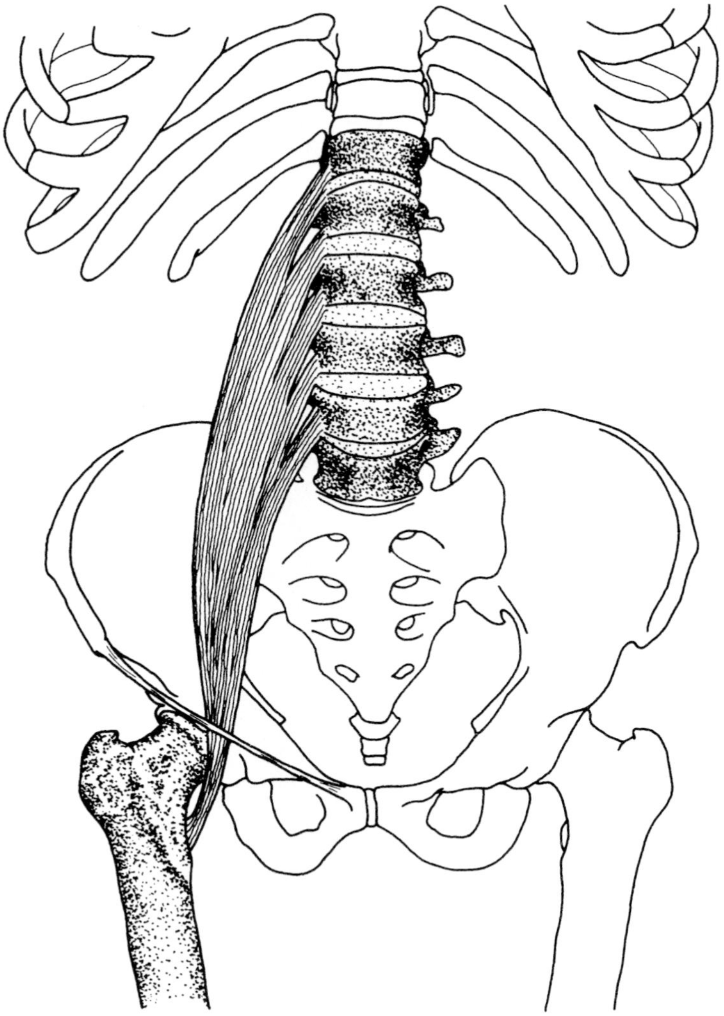

Lumbar region, hip, and thigh—anterior view

Origin Bases of transverse processes of all lumbar vertebrae, bodies of twelfth thoracic and all lumbar vertebrae, intervertebral disks above each lumbar vertebra

Insertion Lesser trochanter of femur

Action Flexes thigh at hip joint, flexes vertebral column

Nerve Branches from lumbar plexus (L2, L3) and sometimes L1 or L4

Note: Some upper fibers insert onto the hip bone from the arcuate line to the iliopectineal eminence to form the *psoas minor.* This muscle has little function and is frequently absent.

ILIACUS

(Part of iliopsoas)

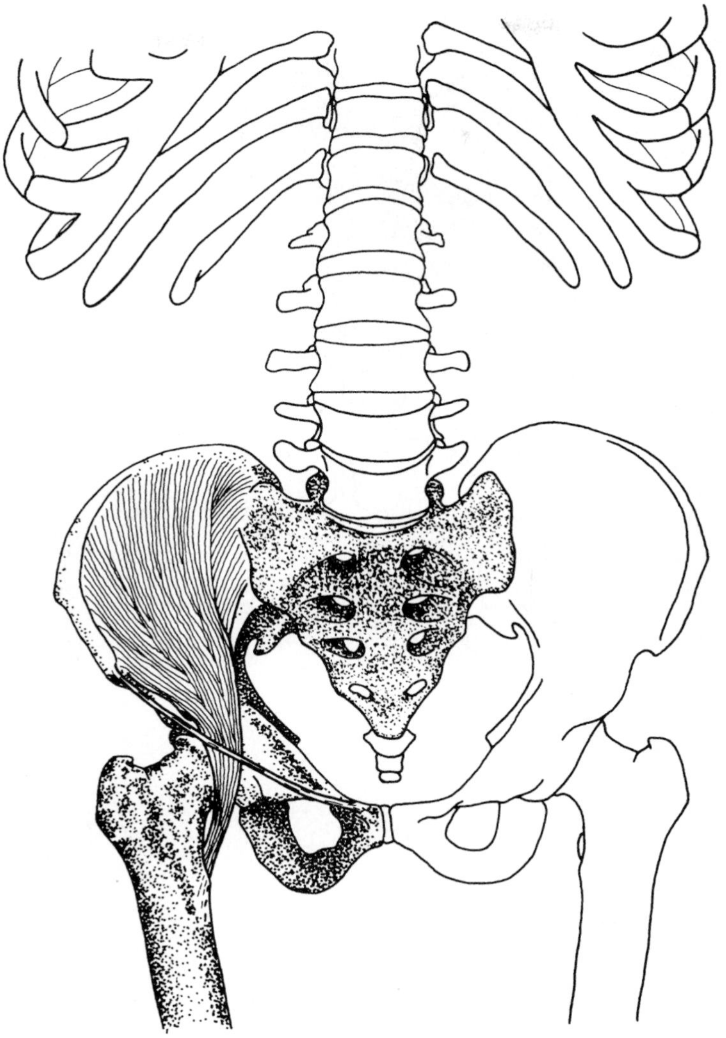

Lumbar region, hip, and thigh—anterior view

Origin Upper two-thirds of iliac fossa, ala of sacrum and adjacent ligaments, anterior inferior iliac spine

Insertion Onto tendon of psoas major, which continues into lesser trochanter of femur (together the two muscles form the iliopsoas)

Action Flexes thigh at hip joint

Nerve Femoral nerve (L2, L3)

Note: The iliacus brings swinging leg forward in walking or running.

PIRIFORMIS

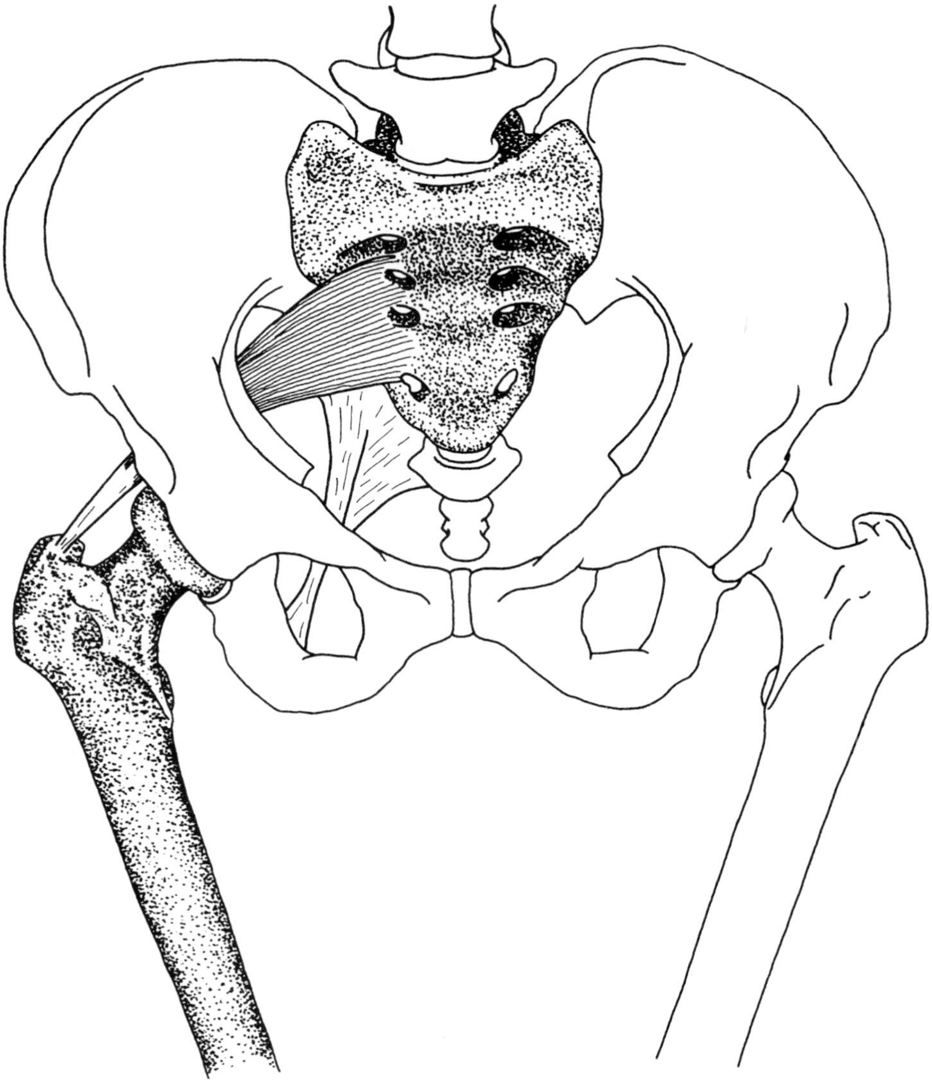

Hip and thigh—anterior view

Origin Internal surface of sacrum, sacrotuberous ligament

Insertion Upper border of greater trochanter

Action Laterally rotates thigh at hip joint, abducts thigh

Nerve Anterior rami of first and second sacral nerves

Note: The common peroneal part of the sciatic nerve may emerge through the belly of the piriformis instead of below its inferior border along with the tibial part.

OBTURATOR INTERNUS

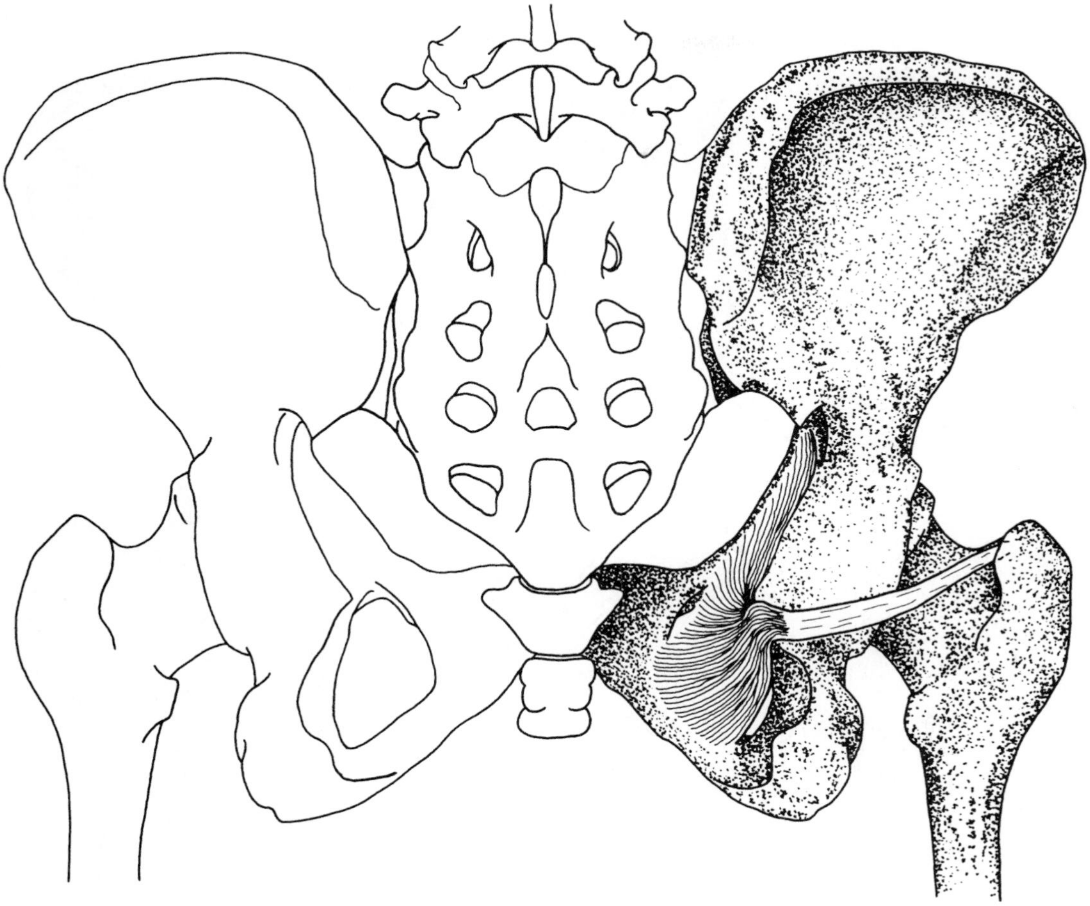

Hip—posterior view

Origin Pelvic surface of obturator membrane and surrounding bones (ilium, ischium, pubis)

Insertion Common tendon with superior and inferior gemelli to medial surface of greater trochanter

Action Laterally rotates thigh at hip joint

Nerve Nerve from sacral plexus (L5, S1–S3)

GEMELLUS SUPERIOR

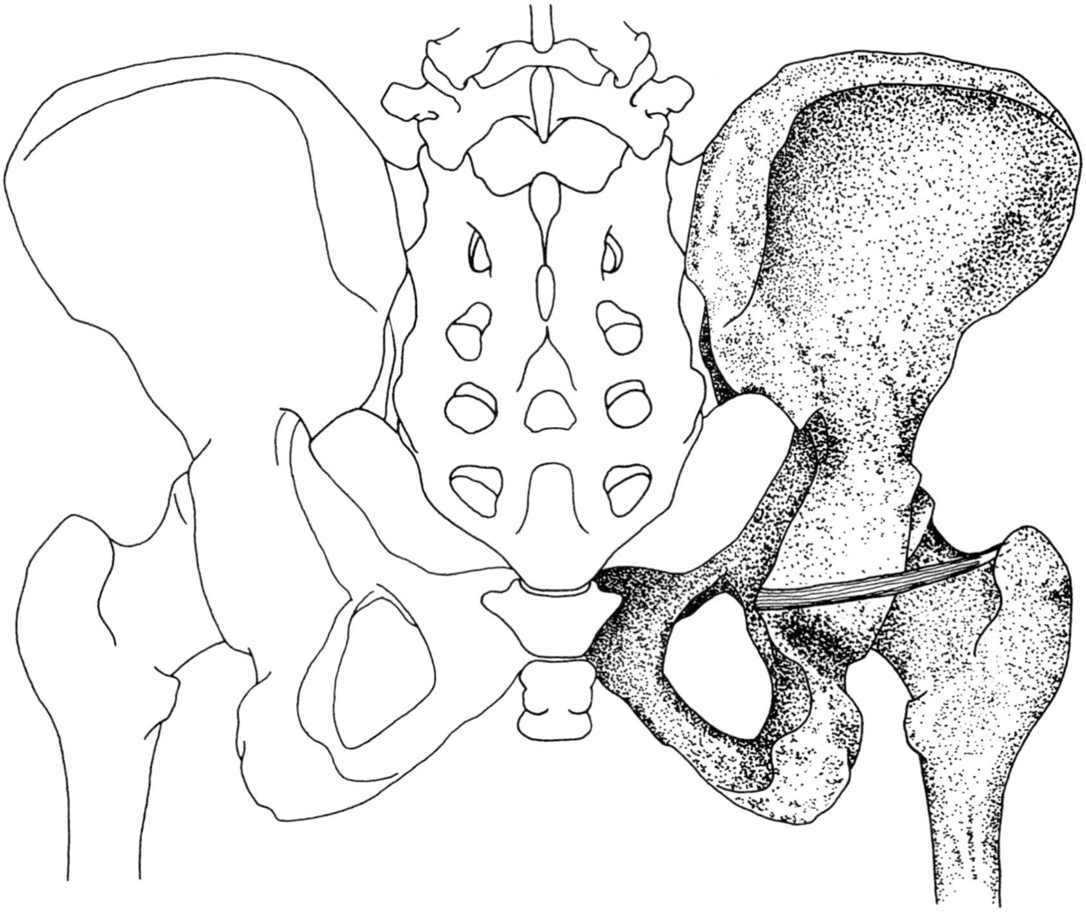

Hip—posterior view

Origin Spine of ischium

Insertion With tendon of obturator internus into upper border of greater trochanter

Action Laterally rotates thigh at hip joint

Nerve Branch of nerve to obturator internus from sacral plexus

GEMELLUS INFERIOR

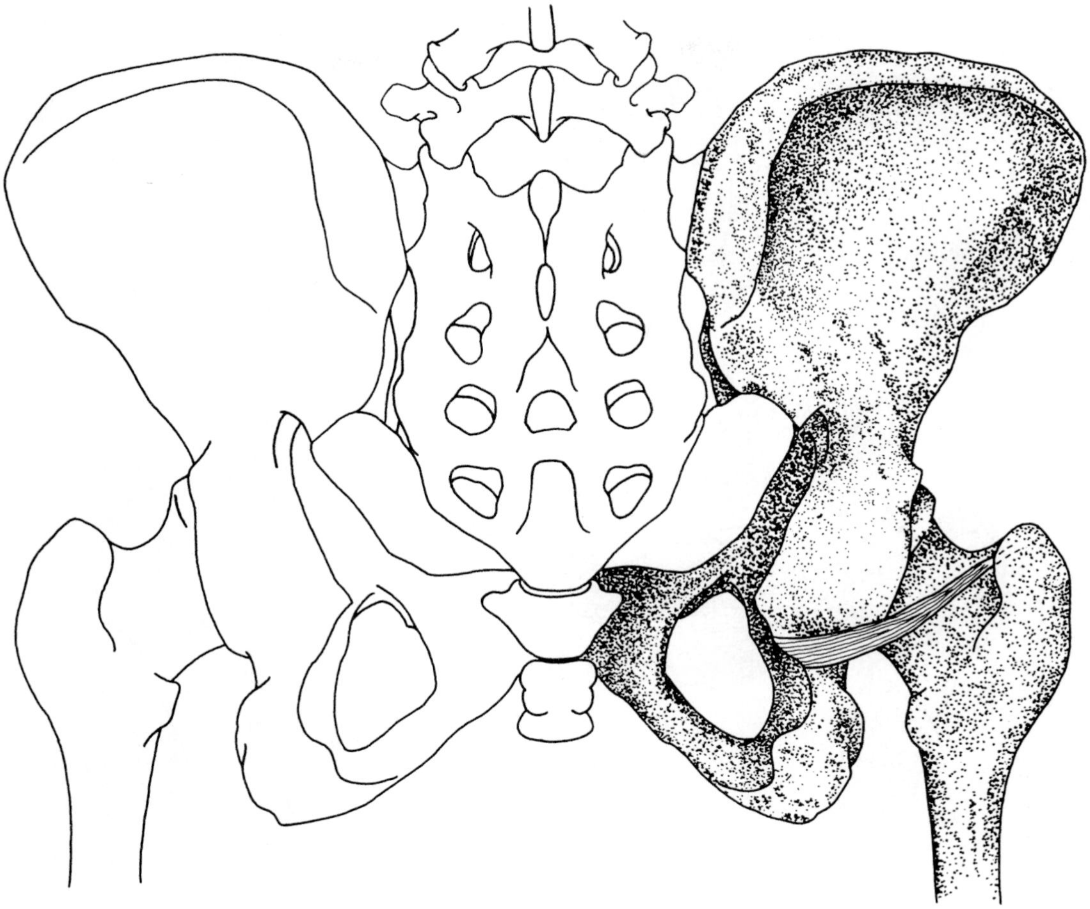

Hip—posterior view

Origin Upper margin of ischial tuberosity

Insertion With tendon of obturator internus into upper border of greater trochanter

Action Laterally rotates thigh at hip joint

Nerve Branch of nerve to quadratus femoris from sacral plexus

OBTURATOR EXTERNUS

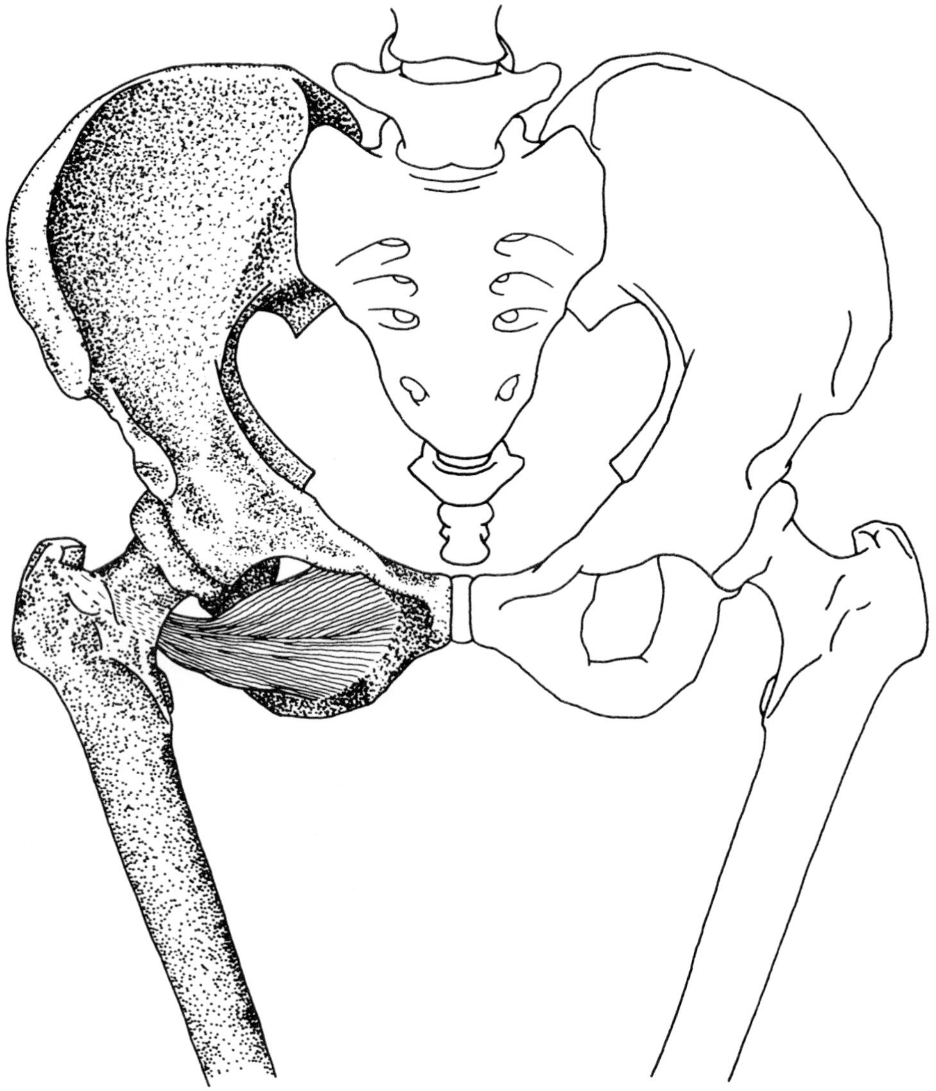

Hip and thigh—anterior view

Origin Outer surface of superior and inferior rami of pubis and ramus of ischium surrounding obturator foramen

Insertion Trochanteric fossa of femur

Action Laterally rotates thigh

Nerve Obturator nerve (L3, L4)

Note: Part of this muscle can be seen posteriorly by separating the gemellus inferior and quadratus femoris. It is deep within this cleft.

QUADRATUS FEMORIS

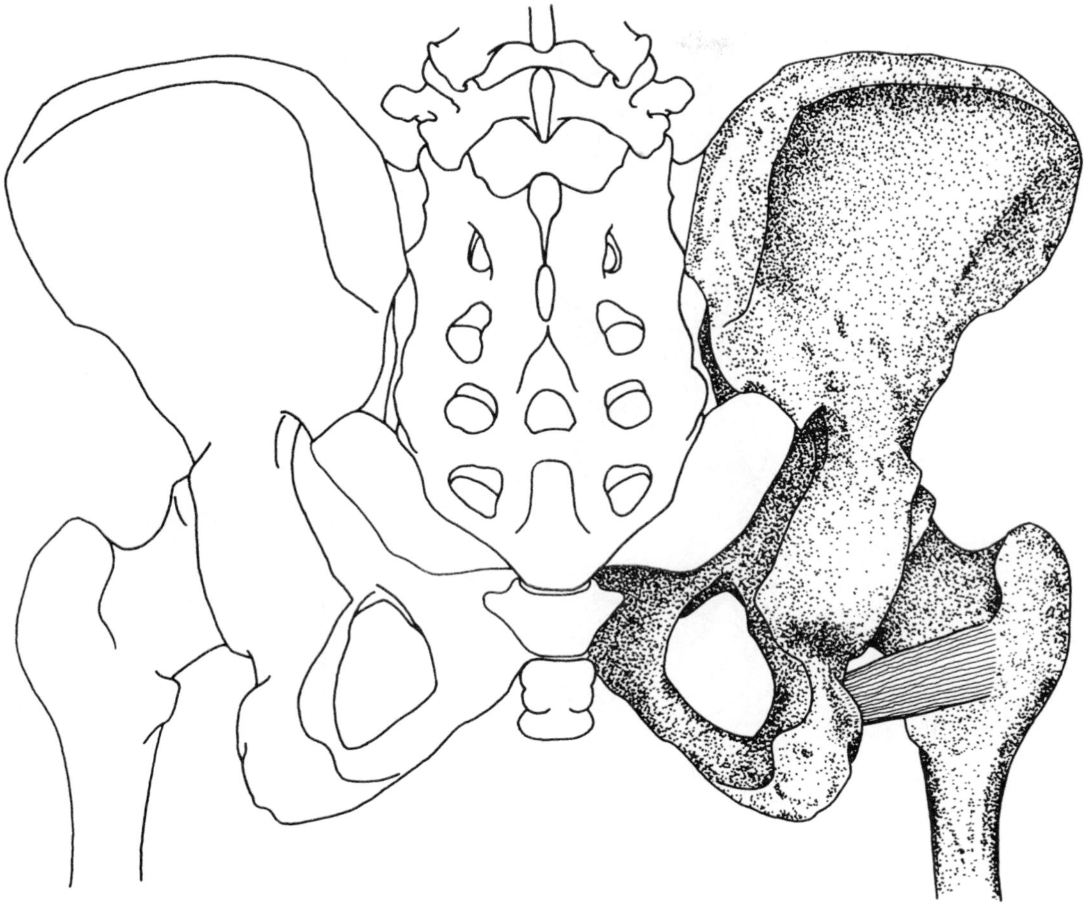

Hip and thigh—posterior view

Origin Lateral border of ischial tuberosity

Insertion Below intertrochanteric crest (quadrate line)

Action Laterally rotates thigh at hip joint

Nerve Branch from sacral plexus (L5, S1)

GLUTEUS MAXIMUS

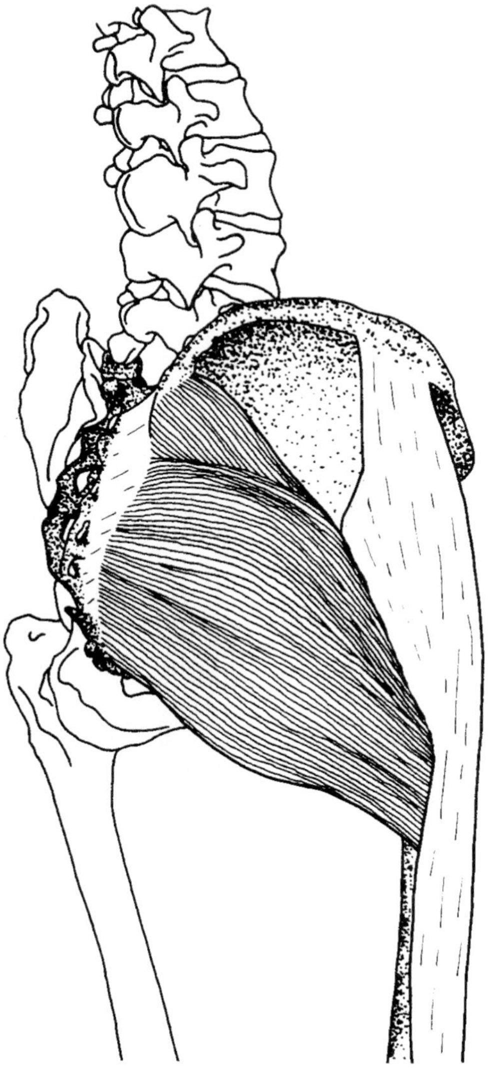

Hip and thigh—lateral view

Origin Outer surface of ilium behind posterior gluteal line, adjacent posterior surface of sacrum and coccyx, sacrotuberous ligament, aponeurosis of erector spinae (sacrospinalis)

Insertion Iliotibial tract of fascia lata, gluteal tuberosity of femur

Action Upper part—abducts, laterally rotates thigh

Lower part—extends, laterally rotates thigh, extends trunk, assists in adduction of thigh

Nerve Inferior gluteal nerve (L5, S1, S2)

Note: This is not a postural muscle; it is not used in walking but only in forceful extension, as in running, climbing, or rising from a seated position.

GLUTEUS MEDIUS

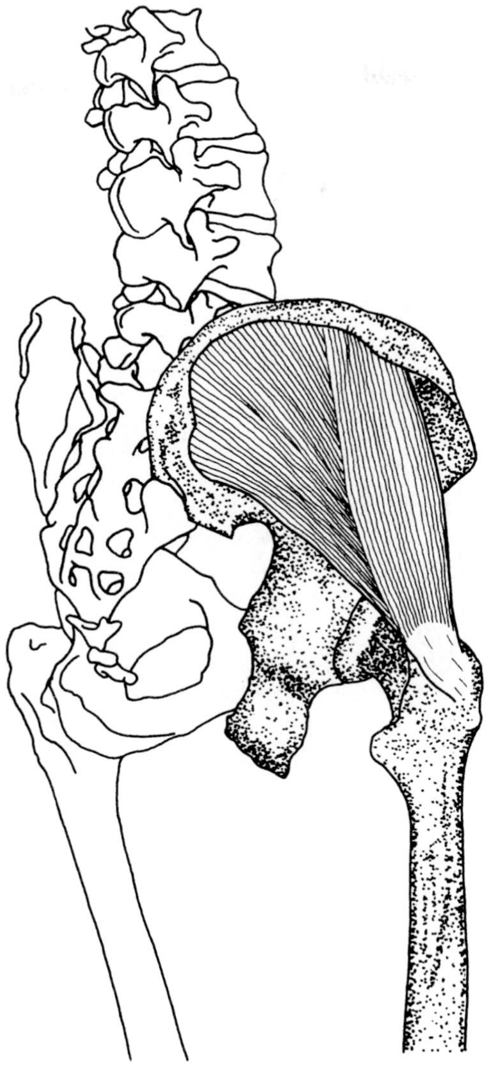

Hip and thigh—lateral view

Origin Outer surface of ilium inferior to iliac crest

Insertion Lateral surface of greater trochanter

Action Abducts femur at hip joint and rotates thigh medially

Nerve Superior gluteal nerve (L4, L5, S1)

Note: In locomotion, this muscle (along with the gluteus minimus) prevents the pelvis from dropping (adduction of thigh) toward the opposite swinging leg.

GLUTEUS MINIMUS

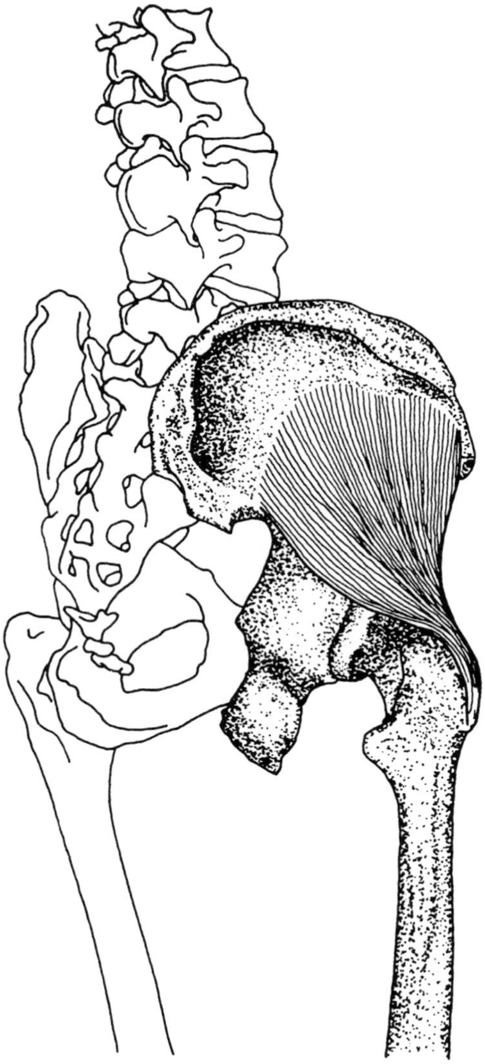

Hip and thigh—lateral view

Origin Outer surface of ilium between middle (anterior) and inferior gluteal lines

Insertion Anterior surface of greater trochanter

Action Abducts femur at hip joint and rotates thigh medially

Nerve Superior gluteal nerve (L4, L5, S1)

Note: See note on gluteus medius.

MUSCLES OF THE HIP

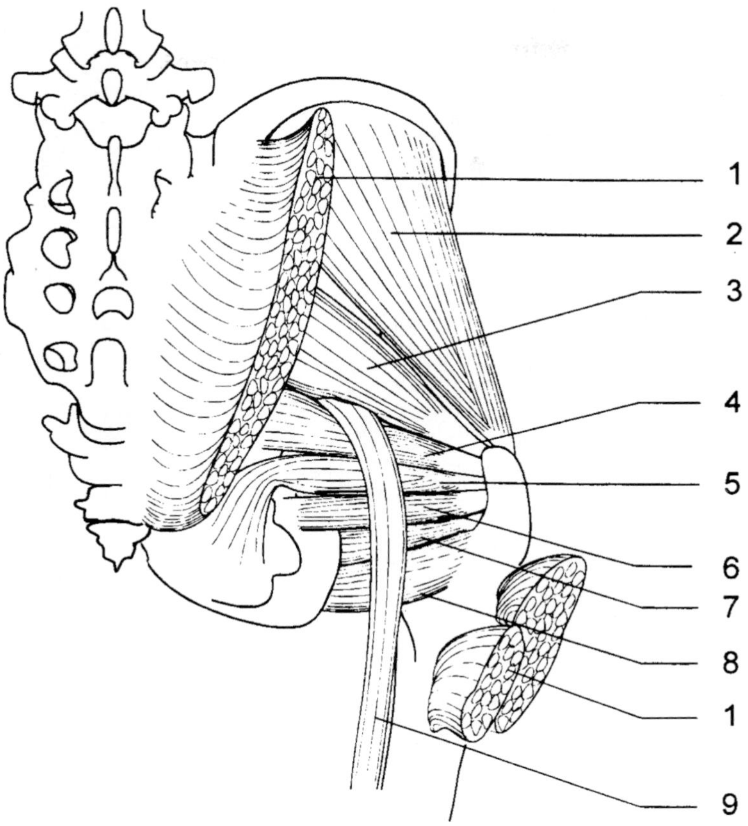

Hip—posterior view

1. Gluteus maximus (cut)
2. Gluteus medius
3. Piriformis
4. Gemellus superior
5. Obturator internus
6. Gemellus inferior
7. Obturator externus
8. Quadratus femoris
9. Sciatic nerve

Note: Gemellus inferior and quadratus femoris have been shown separated to expose the deeply placed obturator externus.

TENSOR FASCIAE LATAE

Origin	Outer edge of iliac crest between anterior superior iliac spine and iliac tubercle
Insertion	Iliotibial tract on upper part of thigh
Action	Flexes, abducts thigh
Nerve	Superior gluteal nerve (L4, L5, S1)

Note: This muscle, along with gluteus maximus, draws the fascia lata upward, stabilizing the knee.

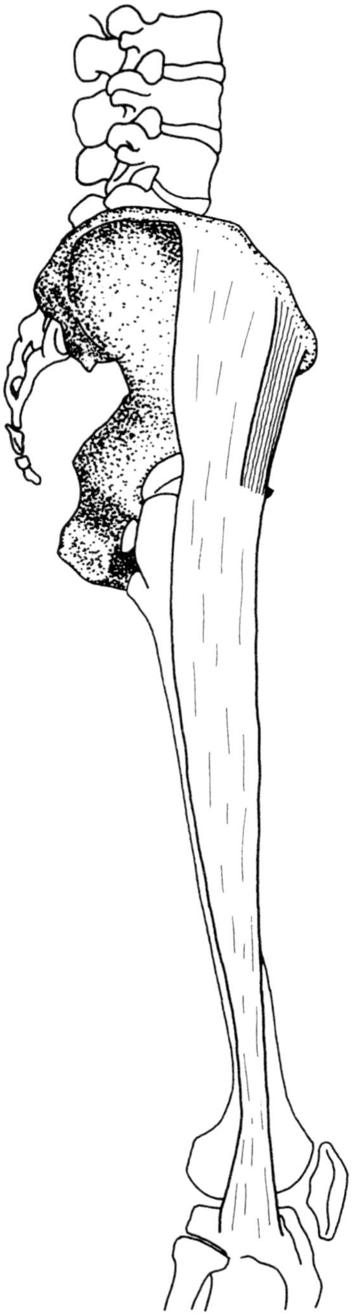

Hip and thigh—lateral view

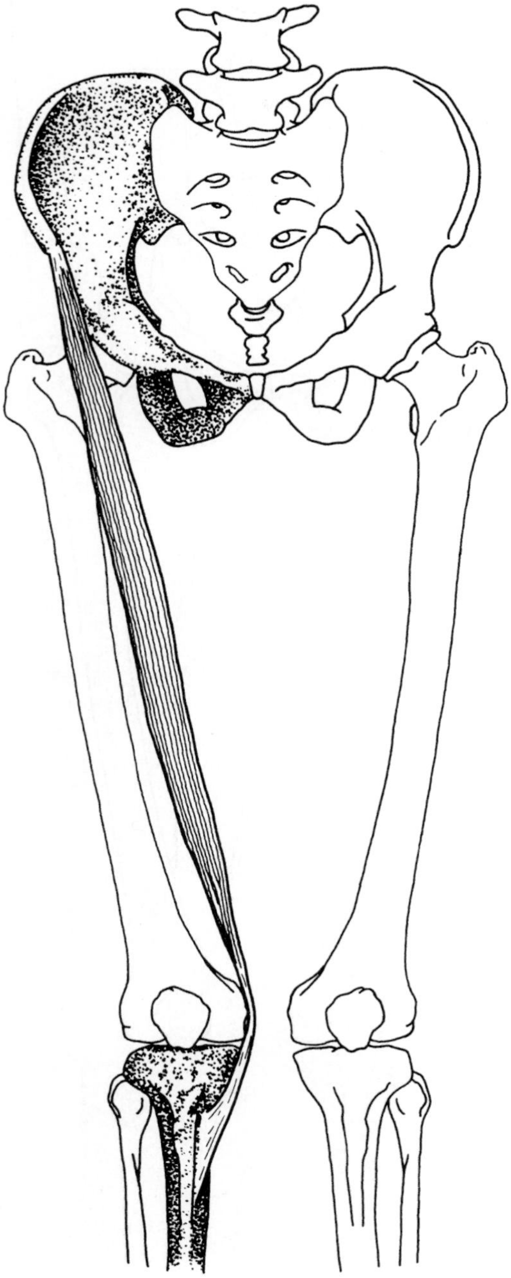

Hip, thigh, and leg—anterior view

SARTORIUS

Origin	Anterior superior iliac spine and area immediately below it
Insertion	Upper part of medial surface of shaft of tibia
Action	Flexes, abducts, and laterally rotates thigh at hip joint, flexes and slightly medially rotates leg at knee joint after flexion
Nerve	Femoral nerve (L2, L3)
Relationships	Insertions of sartorius, gracilis, and semitendinosus fuse on the medial tibia; these tendons, called the pes anserinus (goose foot), give medial support to the knee

Note: This muscle is used to bring swinging leg forward in walking and running.

RECTUS FEMORIS

(One of quadriceps femoris)

Origin	Anterior head—anterior inferior iliac spine Posterior head—ilium above acetabulum
Insertion	Patella, then by patellar ligament to tuberosity of the tibia
Action	Extends leg at knee joint, flexes thigh at hip joint
Nerve	Femoral nerve (L2–L4)

Note: This muscle is used when thigh flexion and leg extension are needed together, such as in kicking a football. In walking, the quadriceps prevent the knee from flexing during heel strike and early support phase.

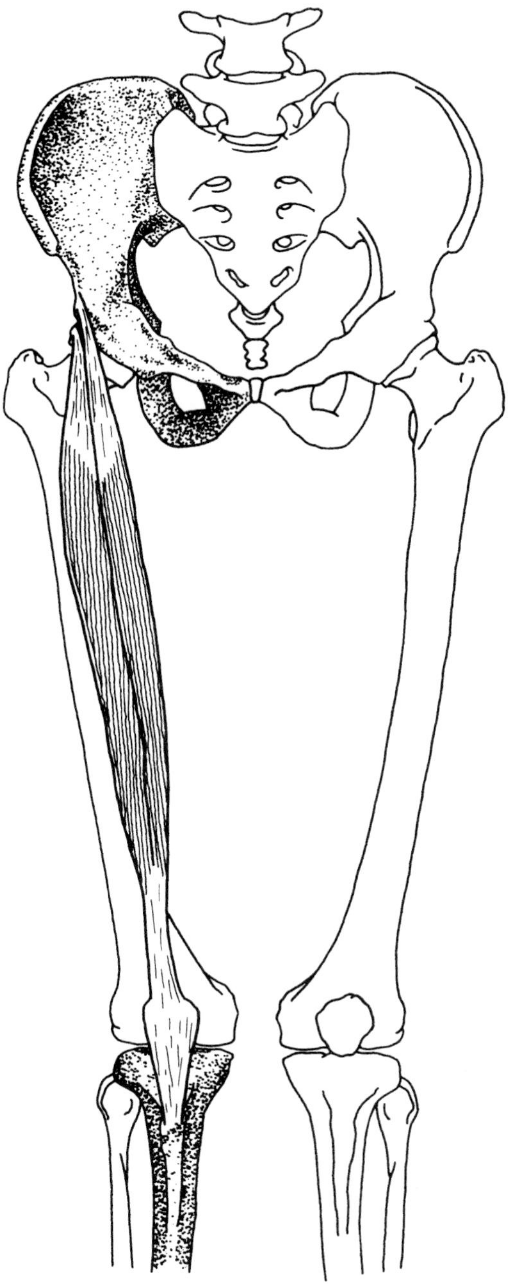

Hip, thigh, and leg—anterior view

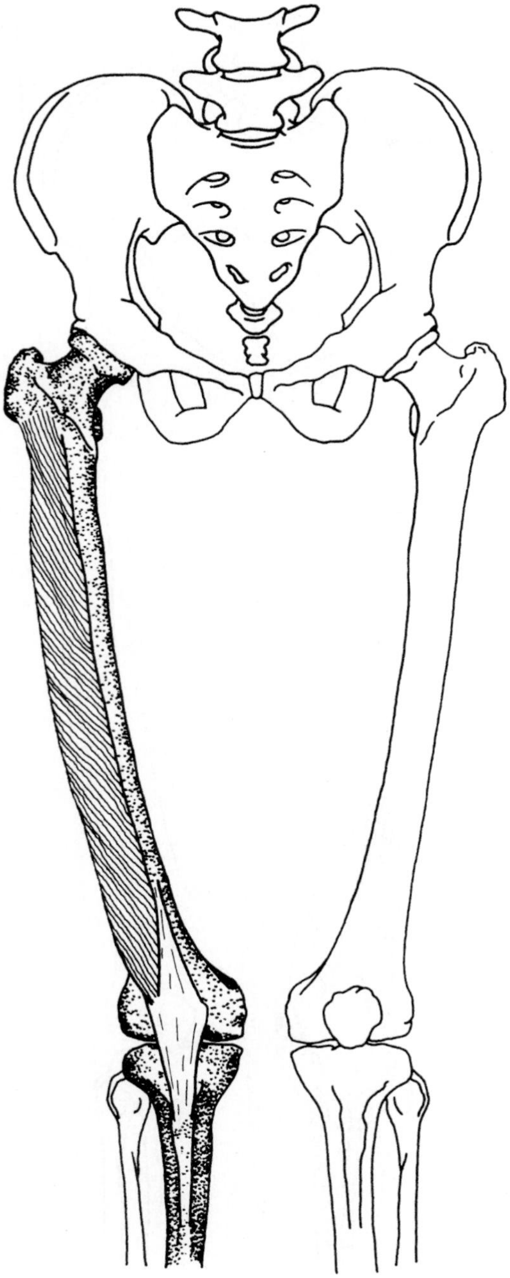

**Hip, thigh, and leg—
anterior view**

VASTUS LATERALIS

(One of quadriceps femoris)

Origin	Intertrochanteric line, inferior border of greater trochanter, gluteal tuberosity, lateral lip of linea aspera of femur
Insertion	Lateral margin of patella, then by patellar ligament to tuberosity of tibia
Action	Extends leg at knee joint
Nerve	Femoral nerve (L2–L4)

VASTUS MEDIALIS

(One of quadriceps femoris)

Origin Intertrochanteric line, medial lip of linea aspera of femur, medial intermuscular septum, medial supracondylar line

Insertion Medial border of the patella, then by patellar ligament into tibial tuberosity, medial condyle of tibia

Action Extends leg at knee joint

Nerve Femoral nerve (L2–L4)

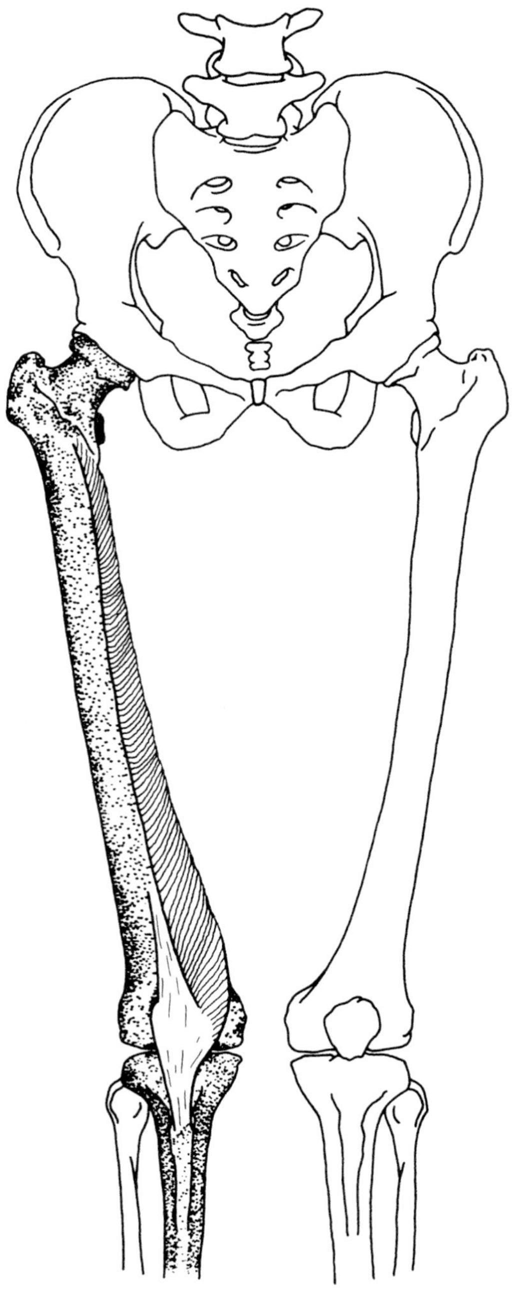

Hip, thigh, and leg—anterior view

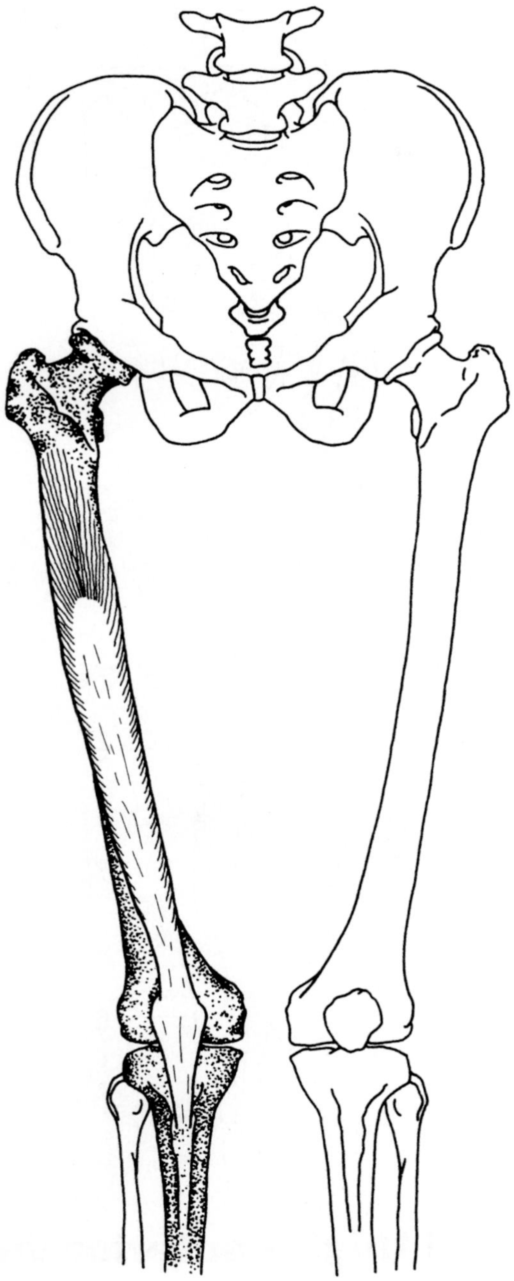

Hip, thigh, and leg—anterior view

VASTUS INTERMEDIUS

(One of quadriceps femoris)

Origin	Anterior and lateral surfaces of upper two-thirds of femur, lateral intermuscular septum, linea aspera, lateral supracondylar line
Insertion	Deep aspect of quadriceps tendon, then through patella to tibial tuberosity
Action	Extends leg at knee joint
Nerve	Femoral nerve (L2–L4)

Note: A few bundles of fibers from this muscle insert onto the upper part of the joint capsule of the knee. They probably draw the capsule superiorly during extension of the leg, preventing it from binding in the joint. They are called *articularis genus.*

MUSCLES OF THE ANTERIOR THIGH

1. Tensor fasciae latae
2. Iliotibial tract
3. Vastus lateralis (quadriceps femoris)
4. Vastus intermedius (quadriceps femoris)
5. Rectus femoris (cut) (quadriceps femoris)
6. Sartorius
7. Vastus medialis (quadriceps femoris)

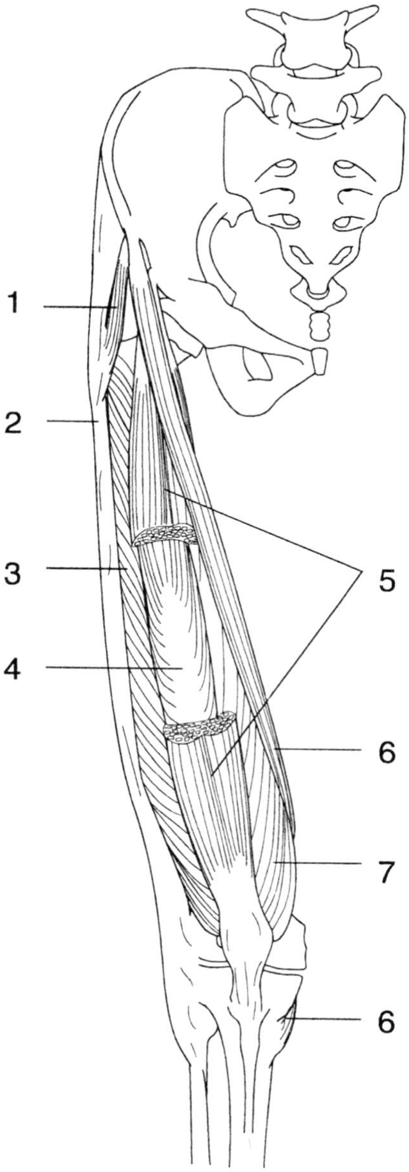

Hip and thigh—anterior view

BICEPS FEMORIS

(Part of hamstrings)

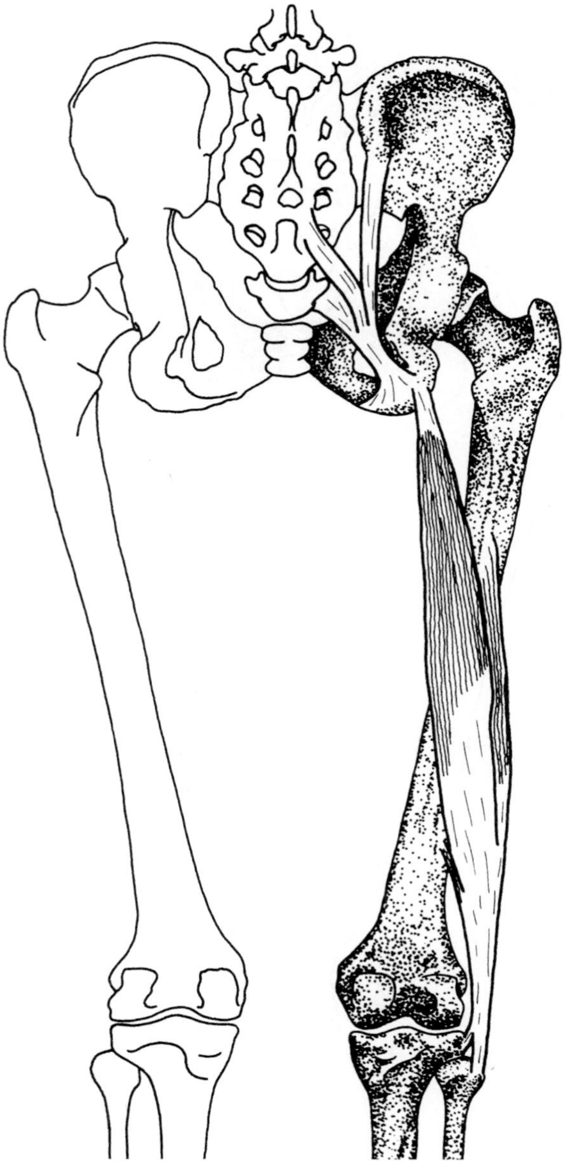

Hip and thigh—posterior view

Origin

Long head—ischial tuberosity, sacrotuberous ligament

Short head—linea aspera, lateral supracondylar ridge, lateral intermuscular septum

Insertion

Lateral side of head of fibula and lateral condyle of tibia

Action

Flexes leg at knee joint, long head also extends thigh at hip joint

Nerve

Long head—tibial part of sciatic nerve (S1–S3)

Short head—common peroneal part of sciatic nerve (L5, S1, S2)

Note: During walking or running, the hamstrings are used to slow down the leg at the end of its swing and prevent the trunk from flexing at the hip. They are susceptible to being strained by resisting the momentum of these body parts.

SEMITENDINOSUS

(Part of hamstrings)

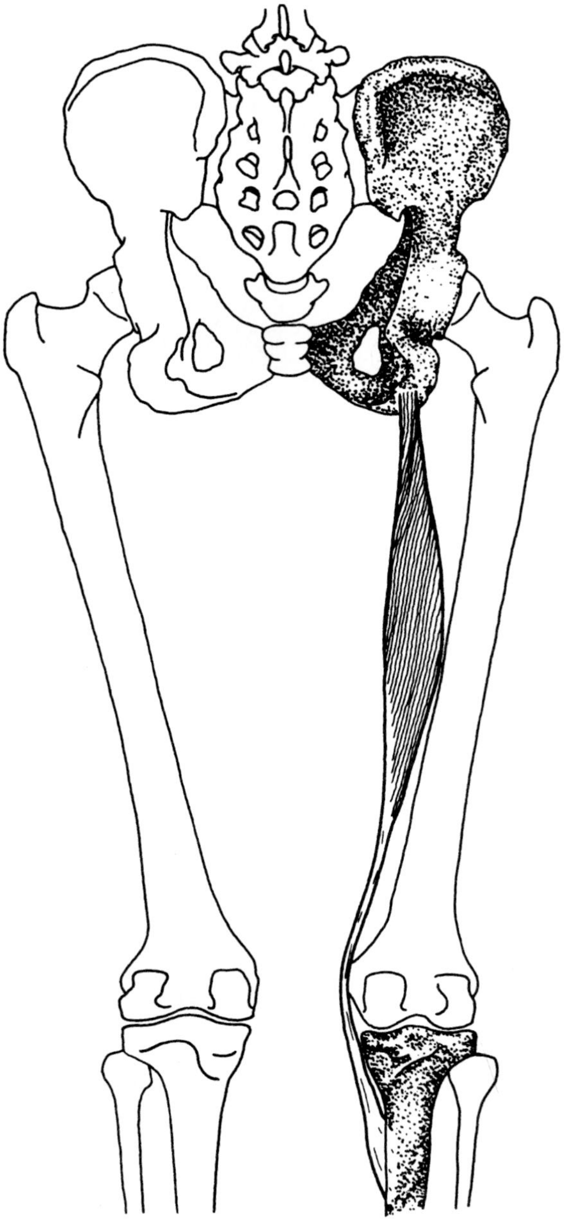

Hip and thigh—posterior view

Origin Ischial tuberosity

Insertion Medial surface of shaft of tibia

Action Flexes and slightly medially rotates leg at knee joint after flexion, extends thigh at hip joint

Nerve Tibial portion of sciatic nerve (L5, S1, S2)

Note: See note on biceps femoris and Relationships section on sartorius.

SEMIMEMBRANOSUS

(Part of hamstrings)

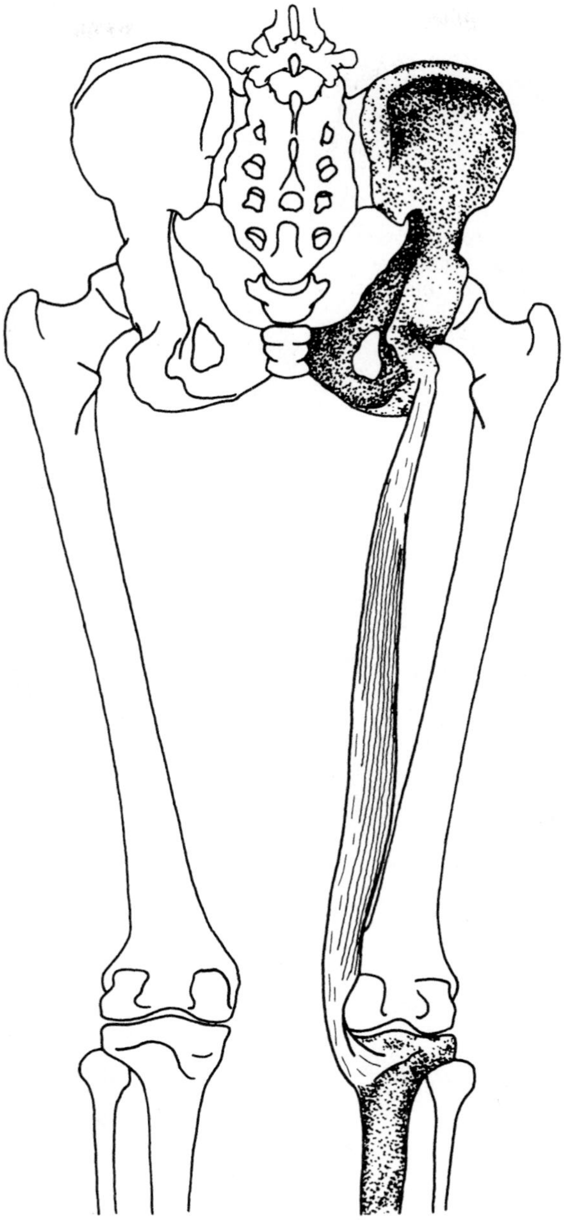

Hip and thigh—posterior view

Origin Ischial tuberosity

Insertion Posterior part of medial condyle of tibia

Action Flexes and slightly medially rotates leg at knee joint after flexion, extends thigh at hip joint

Nerve Tibial portion of sciatic nerve (L5, S1, S2)

Note: See note on biceps femoris.

HAMSTRING MUSCLES

1. Sciatic nerve
2. Quadratus femoris
3. Biceps femoris
4. Semimembranosus
5. Semitendinosus
6. Tibial nerve
7. Common peroneal nerve

Note: The common peroneal nerve is exposed to compression and damage as it passes over the head of the fibula. The quadratus femoris, a lateral rotator, is included for reference.

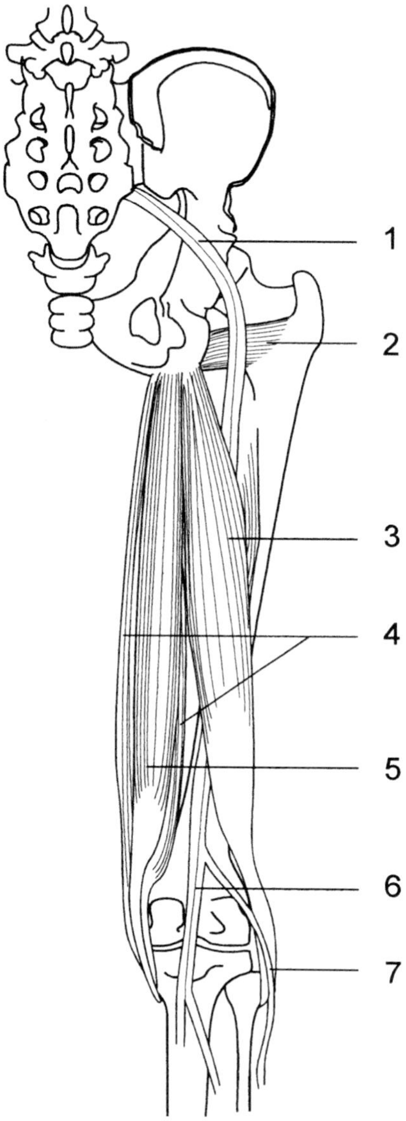

Hip and thigh—posterior view

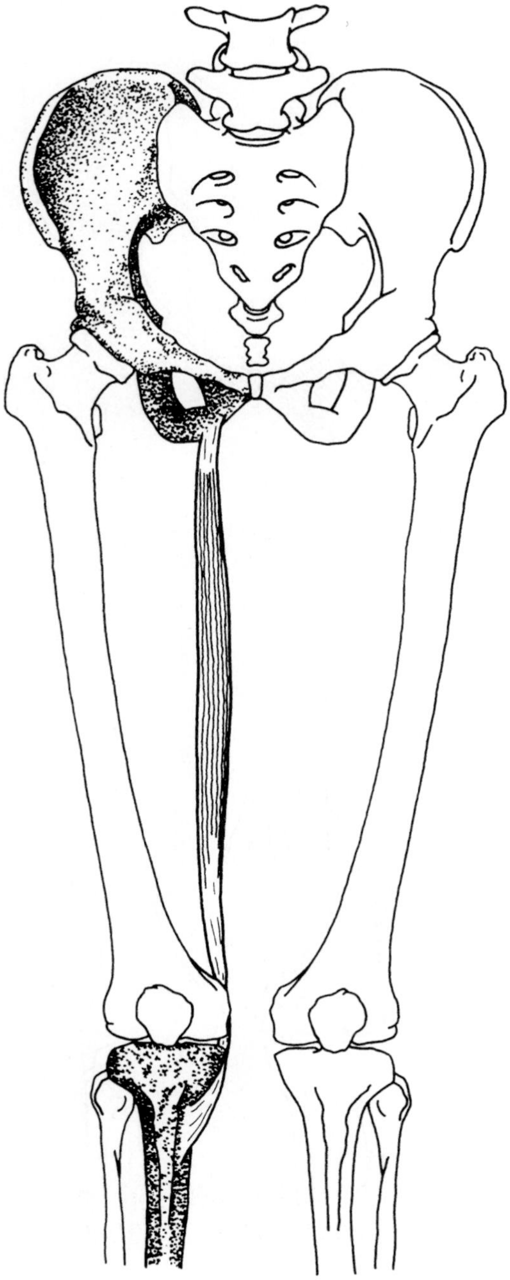

Hip and thigh—anterior view

GRACILIS

Origin	Lower margin of body and inferior ramus of pubis
Insertion	Upper part of medial surface of shaft of tibia
Action	Adducts thigh at hip joint and flexes leg at knee joint; with leg flexed, assists in medial rotation
Nerve	Obturator nerve (L3, L4)

Note: See Relationships section on sartorius.

PECTINEUS

Origin Pectineal line on superior ramus of pubis

Insertion From lesser trochanter to linea aspera of femur

Action Flexes thigh, assists in adduction when hip is flexed

Nerve Femoral nerve (L2–L4), (sometimes a branch of obturator nerve)

Note: The rotating function of this and other hip muscles is controversial and probably depends on whether the hip is flexed or extended and adducted or abducted.

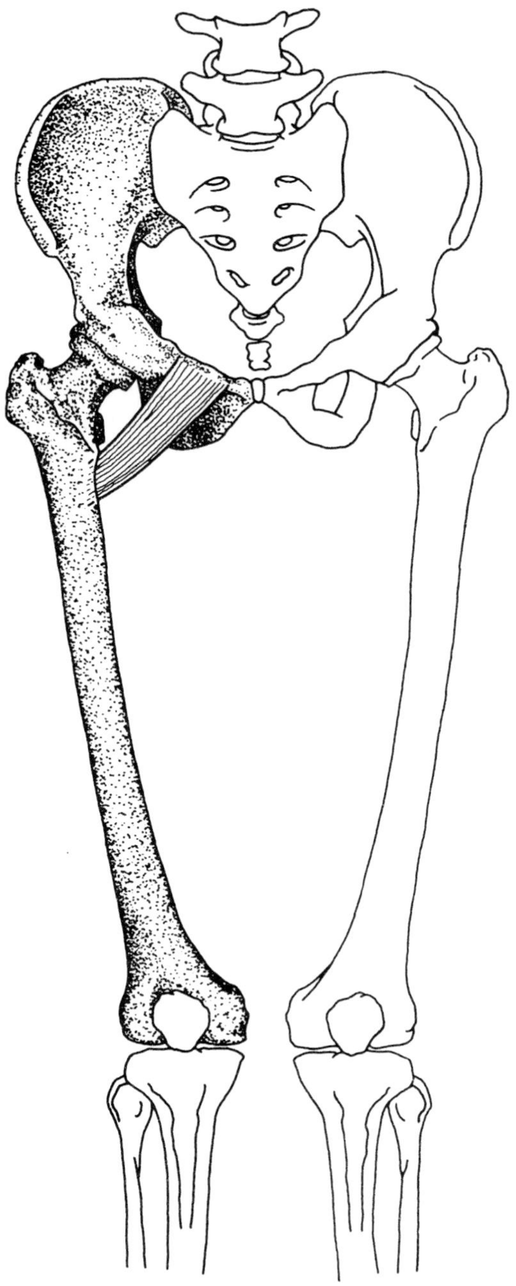

Hip and thigh—anterior view

ADDUCTOR LONGUS

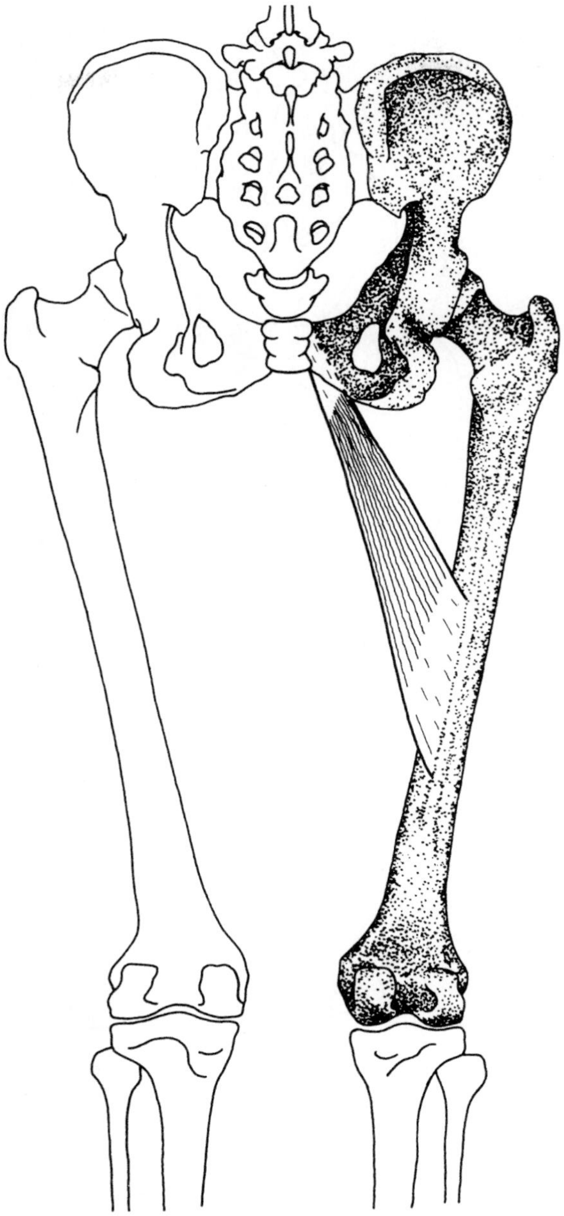

Hip and thigh—posterior view

Origin Anterior of body of pubis

Insertion Medial lip of linea aspera

Action Adducts, flexes thigh, assists in medial rotation

Nerve Obturator nerve (L3, L4)

ADDUCTOR BREVIS

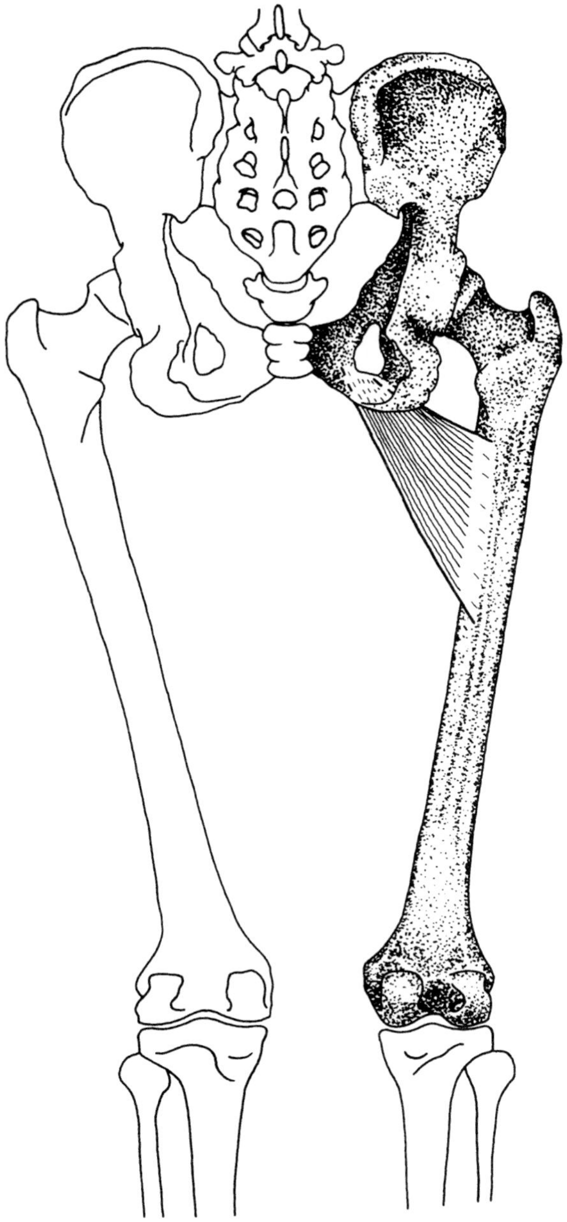

Hip and thigh—posterior view

Origin	Outer surface of inferior ramus of pubis	**Action**	Adducts thigh, assists in flexion, medial rotation
Insertion	From below lesser trochanter to linea aspera and into proximal part of linea aspera	**Nerve**	Obturator nerve (L3, L4)

ADDUCTOR MAGNUS

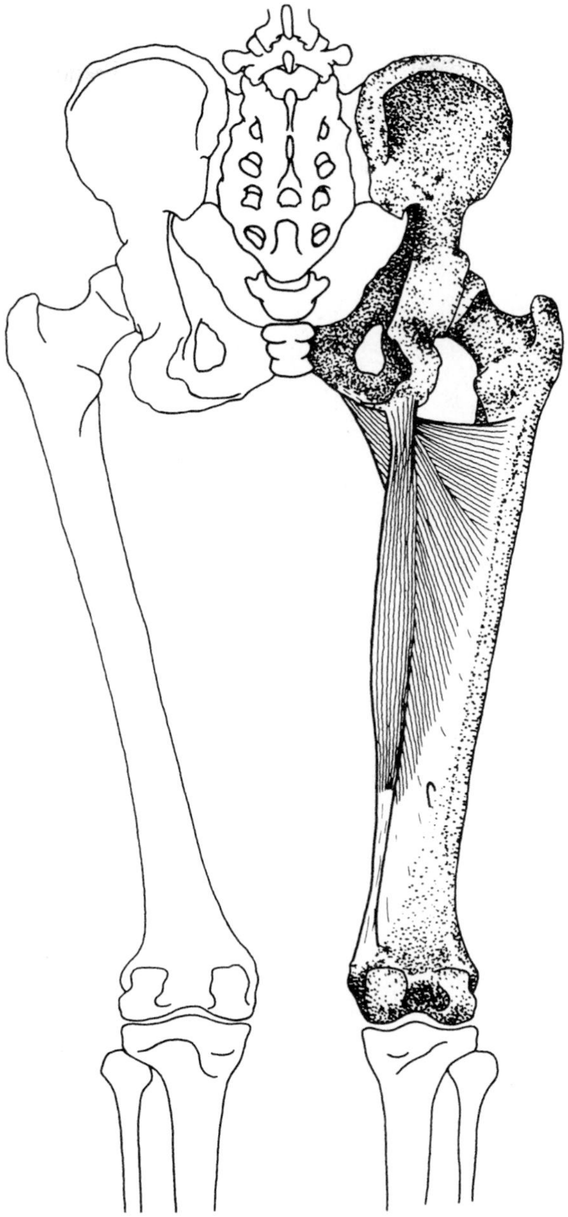

Hip and thigh—posterior view

Origin Inferior ramus of pubis, and ramus and lower part of tuberosity of ischium

Insertion Linea aspera, adductor tubercle of femur

Action Adducts, extends thigh, lower portion (adductor tubercle insertion) assists in medial rotation

Nerve Obturator nerve (L3, L4), sciatic nerve

Note: The linea aspera insertion may assist in lateral rotation.

HIP FLEXORS AND ADDUCTORS

1. Psoas major
2. Iliacus
3. Inguinal ligament
4. Femoral nerve, vein, artery
5. Pectineus
6. Adductor brevis
7. Adductor longus (cut)
8. Adductor magnus
9. Gracilis

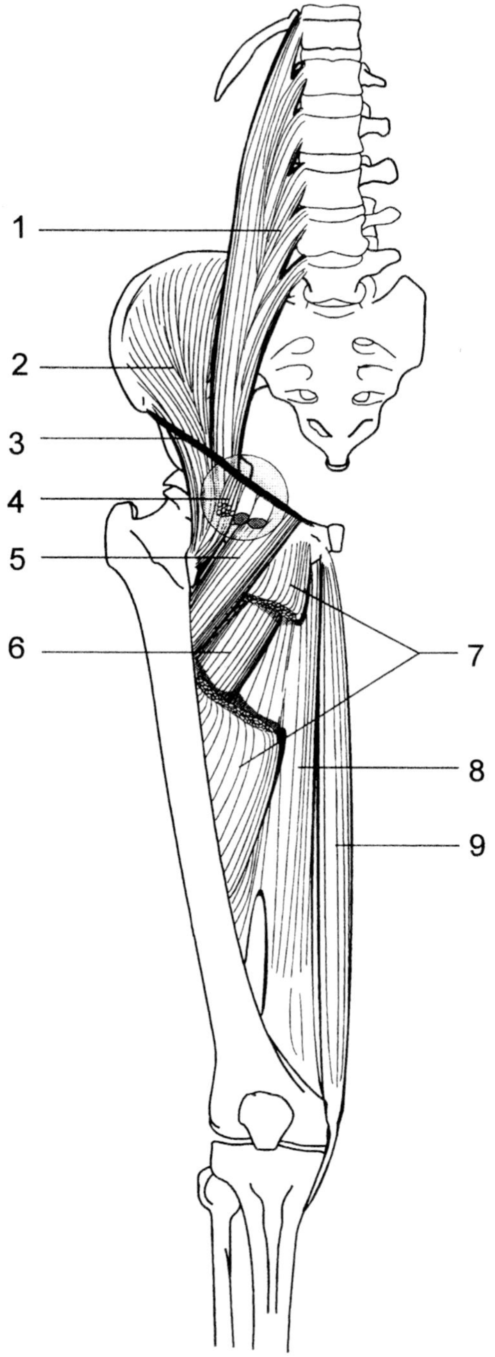

Hip and thigh—anterior view

CHAPTER NINE

MUSCLES OF THE LEG AND FOOT

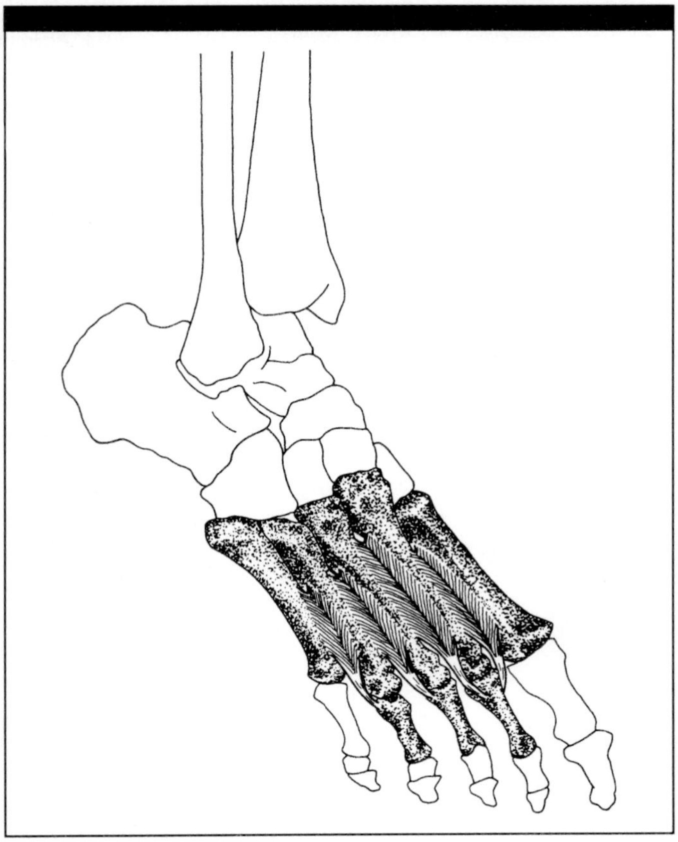

TIBIALIS ANTERIOR

Origin	Lateral condyle of tibia, upper half of lateral surface of tibia, interosseous membrane
Insertion	Medial side and plantar surface of medial cuneiform bone, and base of first metatarsal bone
Action	Dorsiflexes foot at ankle joint, inverts (supinates) foot
Nerve	Deep peroneal nerve (L4, L5, S1)

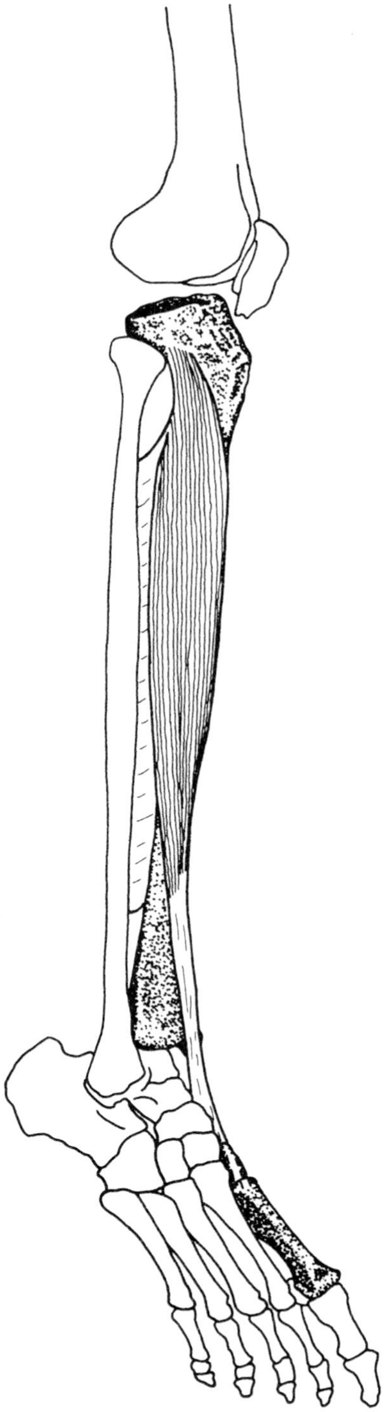

Leg—anterolateral view

EXTENSOR HALLUCIS LONGUS

Origin	Middle half of anterior surface of fibula and interosseous membrane
Insertion	Base of distal phalanx of great toe
Action	Extends, hyperextends great toe, dorsiflexes and inverts (supinates) foot
Nerve	Deep peroneal nerve (L4, L5, S1)

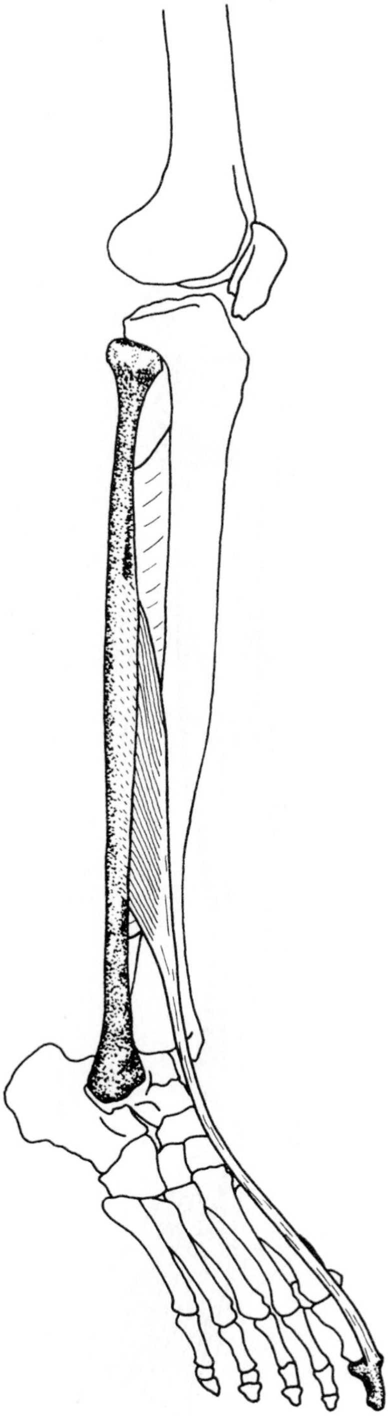

Leg—anterolateral view

EXTENSOR DIGITORUM LONGUS

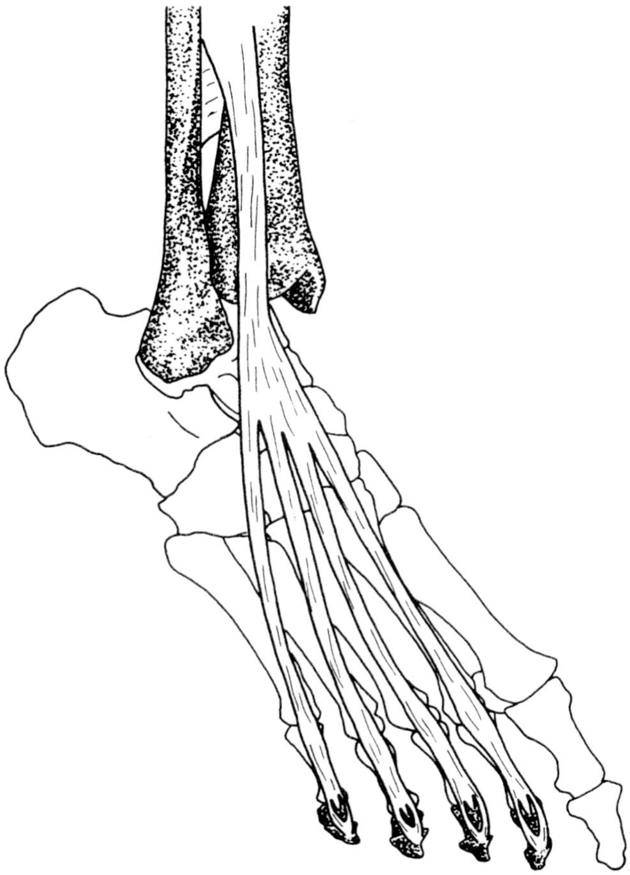

Foot—anterolateral view

Origin	Upper two-thirds of anterior surface of fibula, interosseous membrane, lateral condyle of tibia
Insertion	Along dorsal surface of four lateral toes, and then to bases of middle and distal phalanges
Action	Extends toes, dorsiflexes foot at ankle, everts foot
Nerve	Deep peroneal nerve (L4, L5, S1)

Note: The lower lateral part of this muscle makes a separate insertion onto the dorsal surface of the fifth metatarsal and is called *peroneus tertius.*

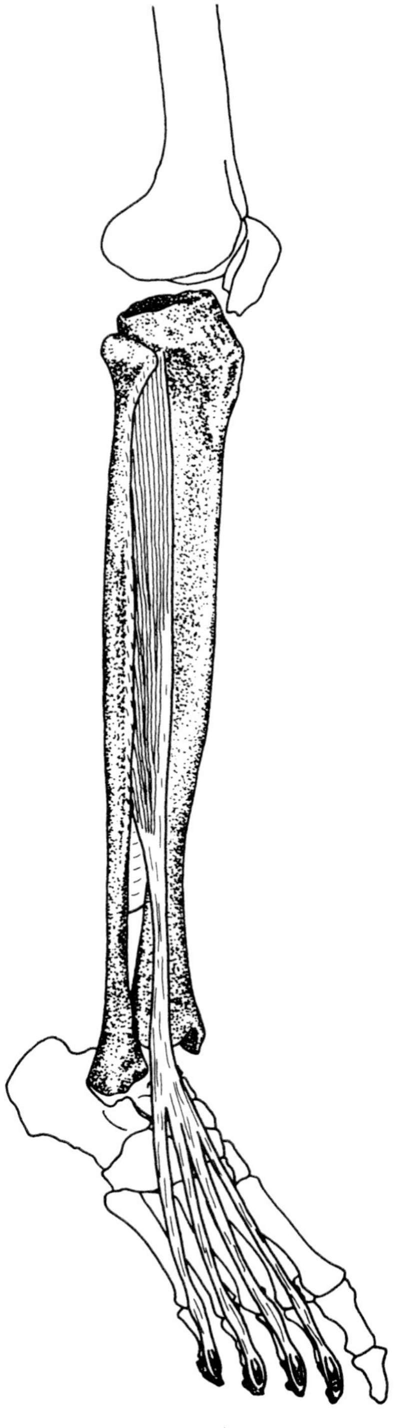

Leg—anterolateral view

PERONEUS TERTIUS

(Lower lateral part of extensor digitorum longus)

Origin	Lower third of anterior surface of fibula and interosseous membrane
Insertion	Dorsal surface of base of fifth metatarsal bone
Action	Dorsiflexes and everts foot
Nerve	Deep peroneal nerve (L4, L5, S1)

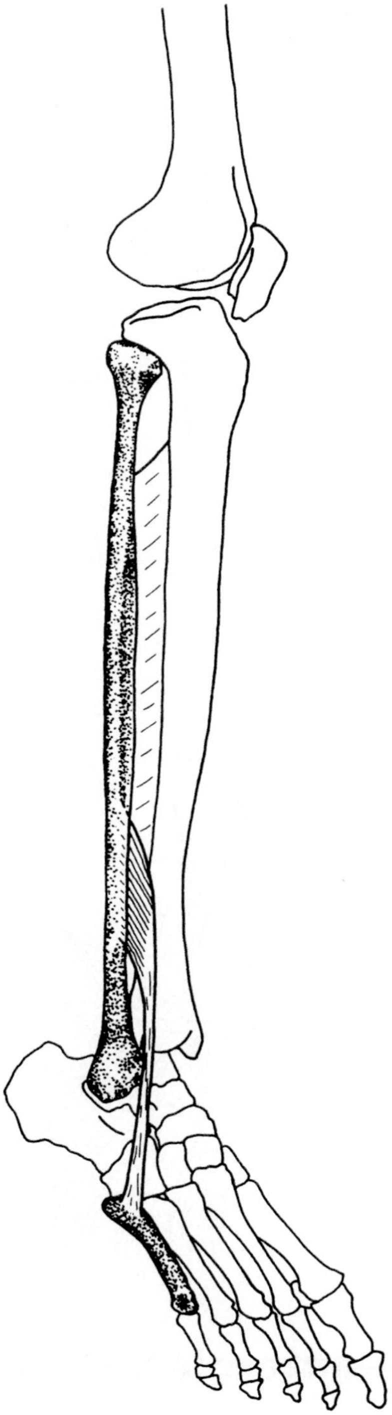

Leg—anterolateral view

ANTERIOR AND LATERAL LEG MUSCLES

1. Peroneus longus
2. Peroneus brevis
3. Peroneus tertius
4. Tibialis anterior
5. Extensor retinaculum
6. Extensor hallucis longus
7. Extensor digitorum longus
8. Extensor digitorum brevis

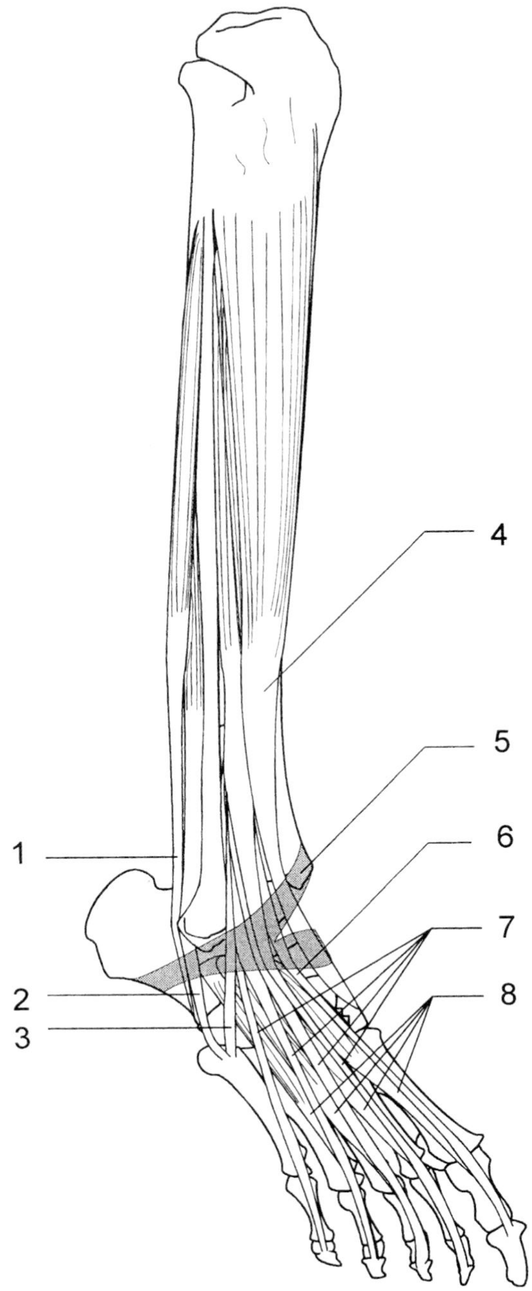

Leg—anterolateral view

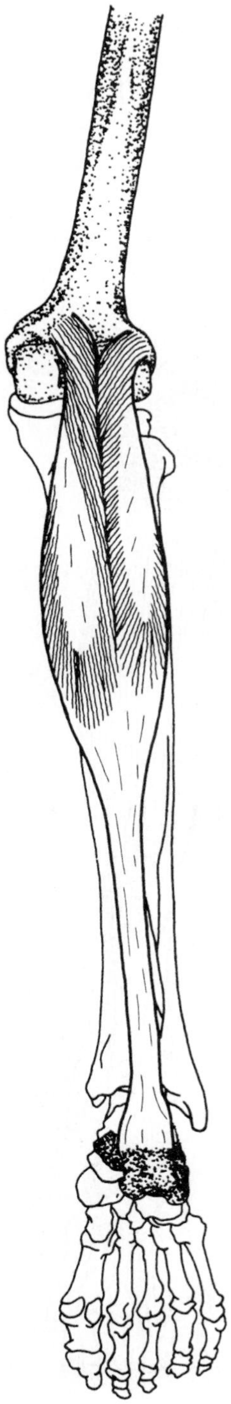

Leg—posterior view

GASTROCNEMIUS

(Part of triceps surae)

Origin	Lateral head—lateral condyle and posterior surface of femur Medial head—popliteal surface of femur above medial condyle
Insertion	Posterior surface of the calcaneus
Action	Plantar flexes foot, flexes leg at knee
Nerve	Tibial nerve (S1, S2)

SOLEUS

(Part of triceps surae)

Origin Posterior surface of the tibia (soleal line), upper third of posterior surface of fibula, fibrous arch between tibia and fibula

Insertion Posterior surface of the calcaneus

Action Plantar flexes foot

Nerve Tibial nerve (S1, S2)

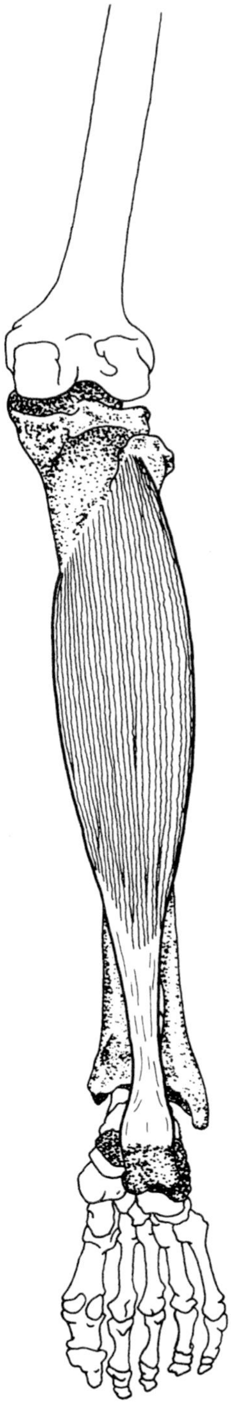

Leg—posterior view

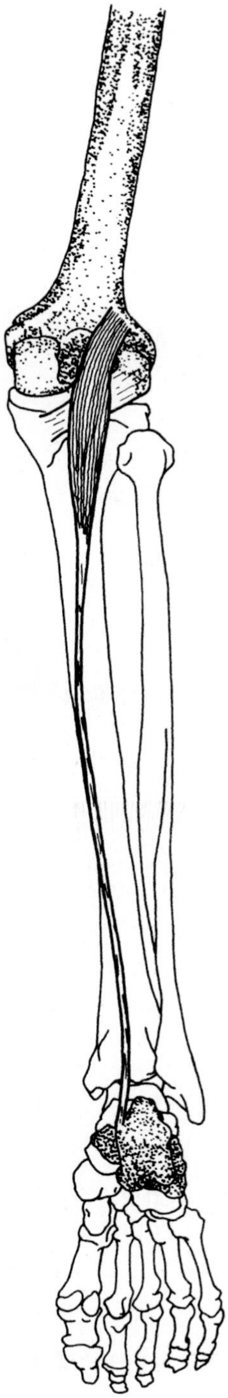

Leg—posterior view

PLANTARIS

Origin	Lateral supracondylar ridge of femur, oblique popliteal ligament
Insertion	Posterior surface of the calcaneus
Action	Plantar flexes foot, flexes leg
Nerve	Tibial nerve (L4, L5, S1)

MUSCLES OF THE CALF

1. Soleus
2. Plantaris
3. Gastrocnemius (cut)

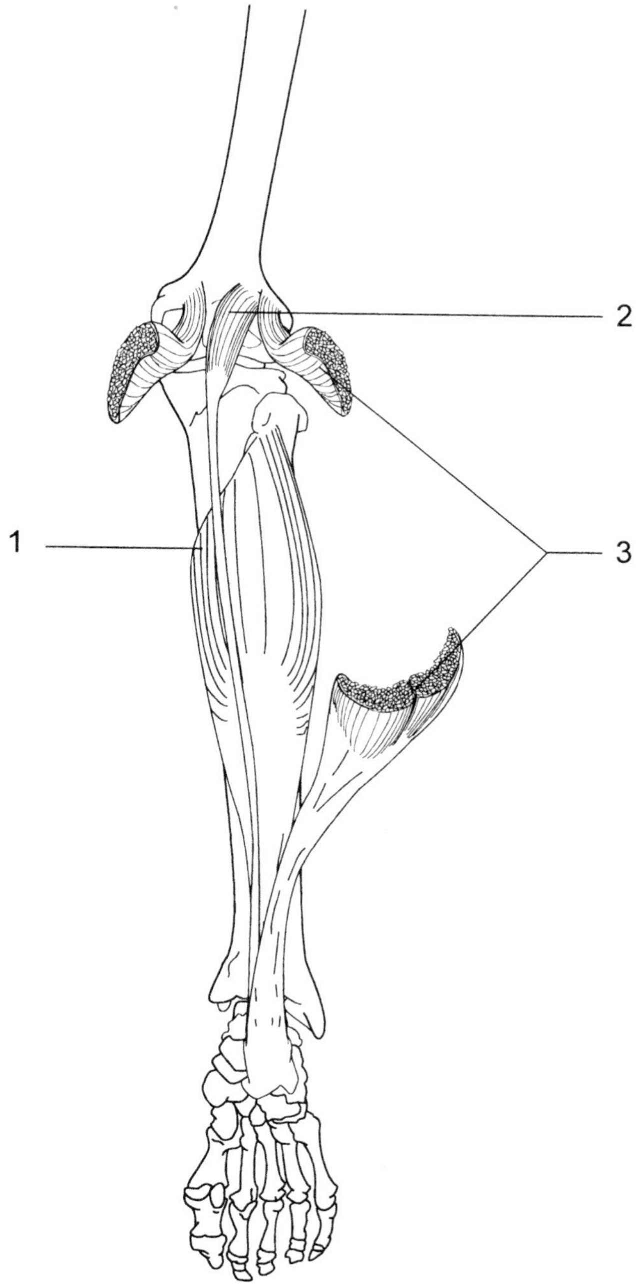

Leg—posterior view

POPLITEUS

Origin	Lateral surface of lateral condyle of femur
Insertion	Upper part of posterior surface of tibia
Action	Rotates leg medially, flexes leg
Nerve	Tibial nerve (L4, L5, S1)

Note: Stern contends that this muscle stabilizes the knee by preventing lateral rotation of the tibia during medial rotation of the thigh while the foot is planted.

Reference:
Stern, JT: *Essentials of Gross Anatomy,* F. A. Davis Company, Philadelphia, 1988.

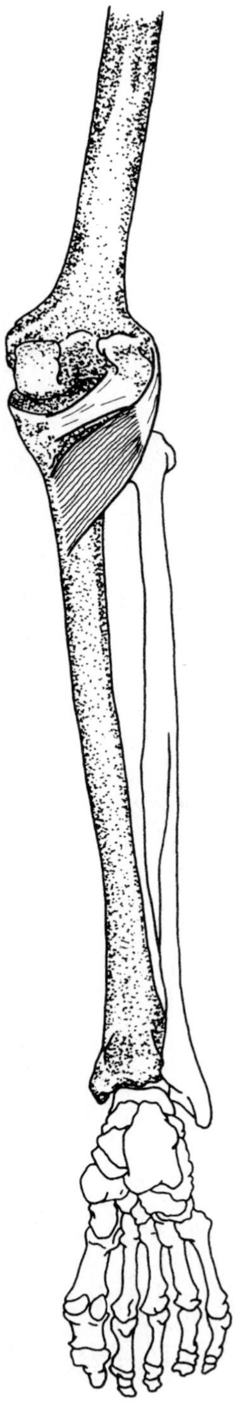

Leg—posterior view

FLEXOR HALLUCIS LONGUS

Origin	Lower two-thirds of posterior surface of shaft of fibula, posterior intermuscular septum, interosseous membrane
Insertion	Base of distal phalanx of great toe
Action	Flexes distal phalanx of great toe, assists in plantar flexing foot, inverts foot
Nerve	Tibial nerve (L5, S1, S2)

Note: This muscle is important in pushing off the surface in walking, running, jumping.

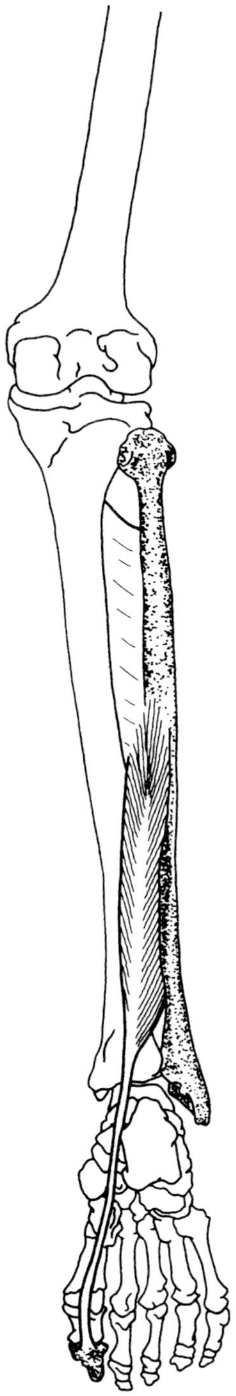

Leg—posterior view

FLEXOR DIGITORUM LONGUS

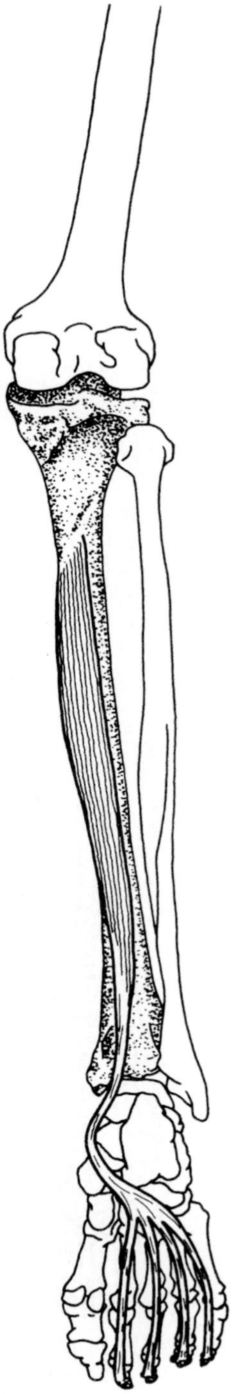

Leg—posterior view

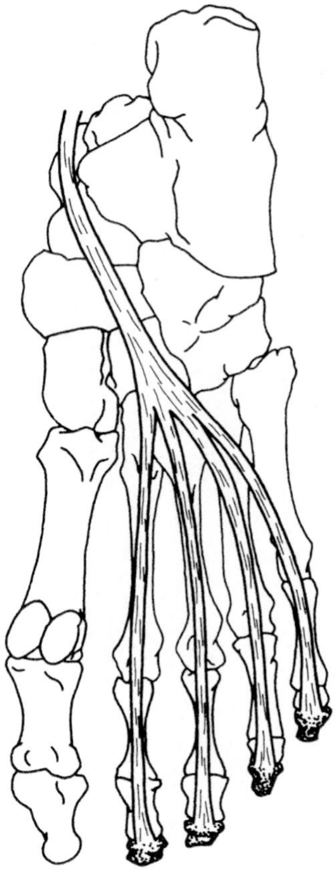

Foot—plantar view

Origin	Medial part of posterior surface of tibia
Insertion	Bases of distal phalanges of second, third, fourth, and fifth toes
Action	Flexes distal phalanges of lateral four toes, assists in plantar flexing foot, inverts foot
Nerve	Tibial nerve (L5, S1)

TIBIALIS POSTERIOR

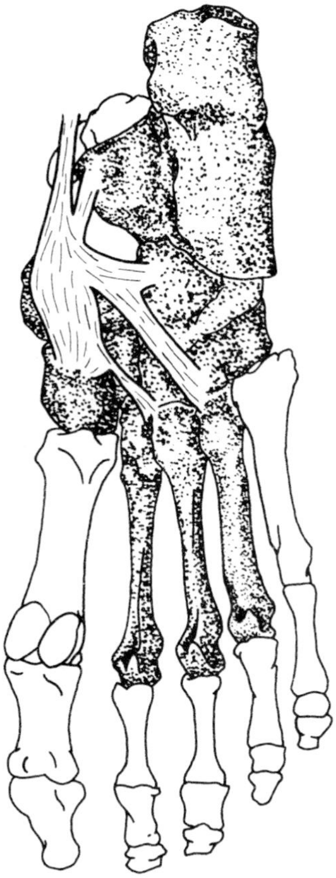

Foot—plantar view

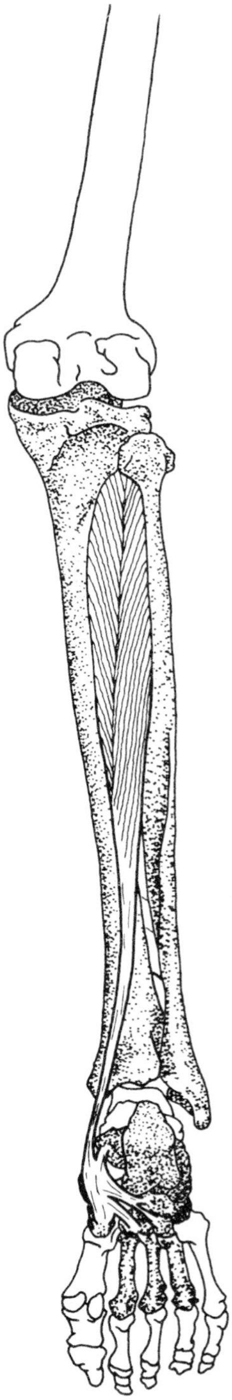

Leg—posterior view

Origin	Lateral part of posterior surface of tibia, interosseous membrane, proximal half of posterior surface of fibula
Insertion	Tuberosity of navicular bone, cuboid, cuneiforms, second, third, and fourth metatarsals, sustentaculum tali of calcaneus
Action	Plantar flexes, inverts foot
Nerve	Tibial nerve (L5, S1)

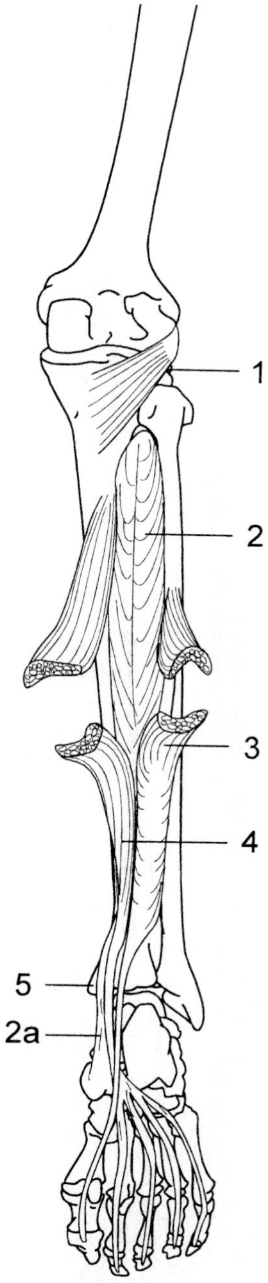

DEEP POSTERIOR LEG MUSCLES

1. Popliteus
2. Tibialis posterior
2a. Tendon of tibialis posterior
3. Flexor hallucis longus (cut)
4. Flexor digitorum longus (cut)
5. Medial malleolus

Leg—posterior view

PERONEUS LONGUS

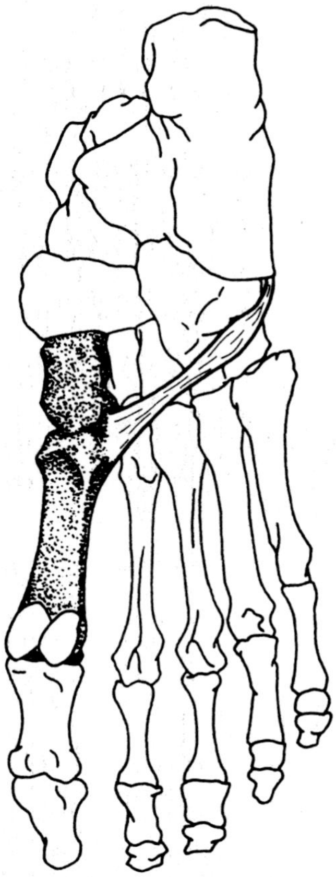

Foot—plantar view

Origin	Upper two-thirds of lateral surface of fibula
Insertion	Lateral side of medial cuneiform, base of first metatarsal
Action	Plantar flexes, everts foot
Nerve	Superficial peroneal nerve (L4, L5, S1)

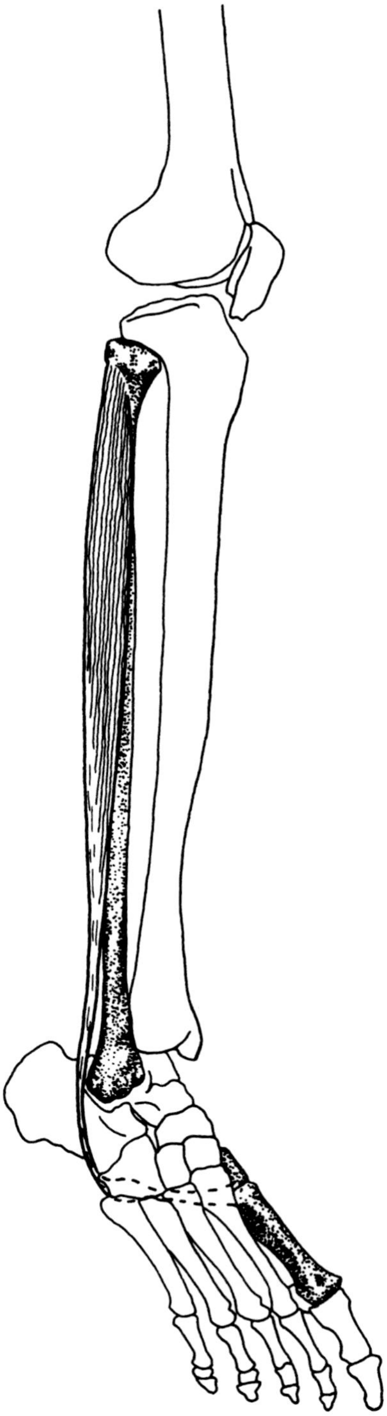

Leg—anterolateral view

PERONEUS BREVIS

Origin	Lower two-thirds of lateral surface of fibula
Insertion	Lateral side of base of fifth metatarsal bone
Action	Everts, plantar flexes foot
Nerve	Superficial peroneal nerve (L4, L5, S1)

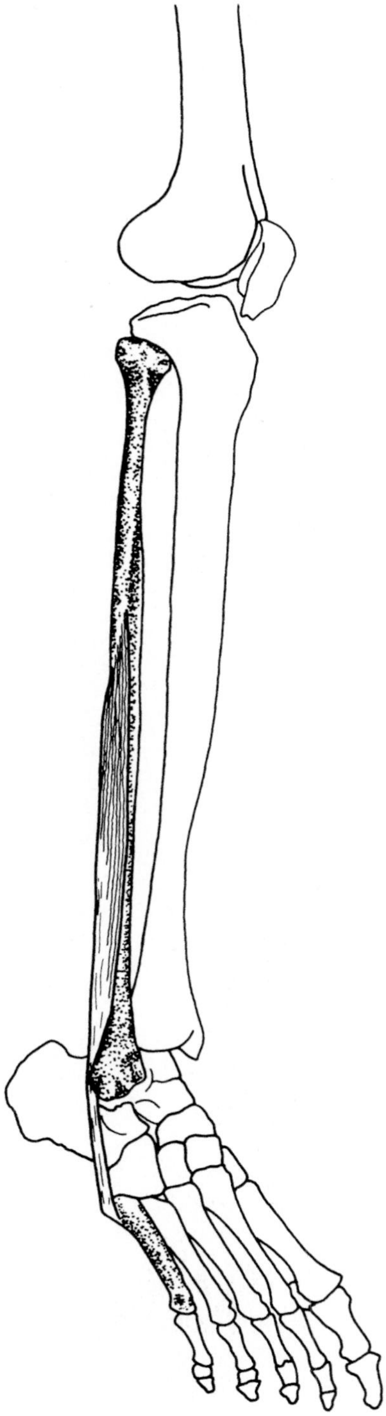

Leg—anterolateral view

EXTENSOR DIGITORUM BREVIS

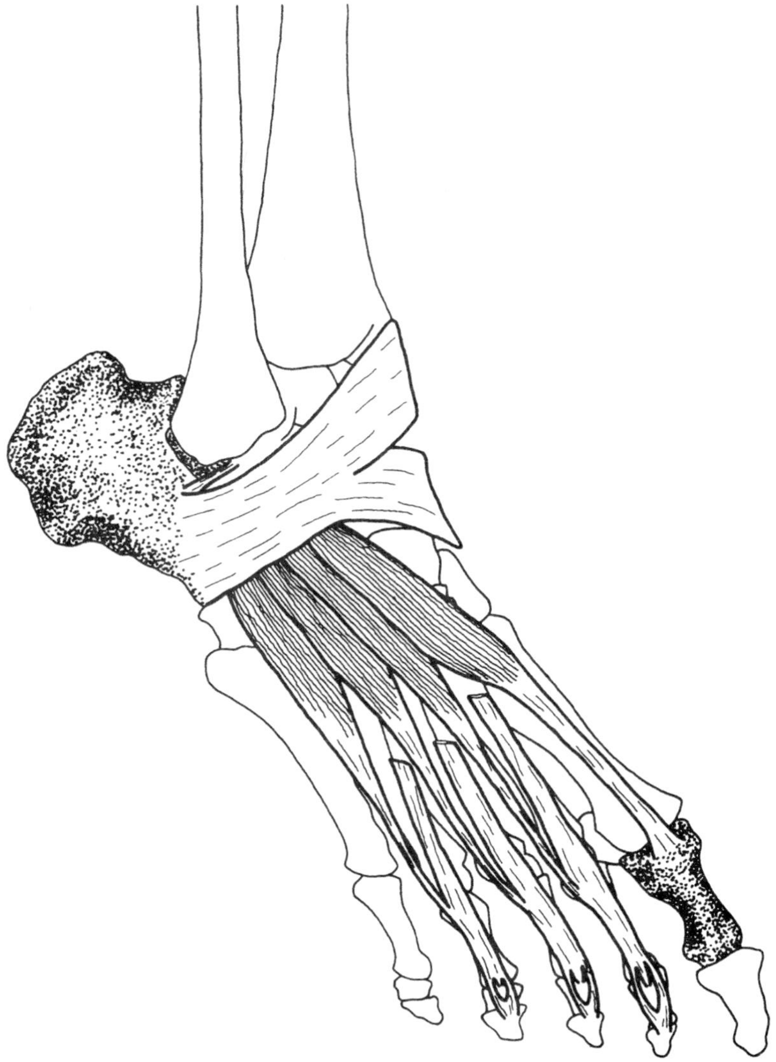

Foot—anterolateral view

Origin Anterior and lateral surfaces of calcaneus, lateral talocalcaneal ligament, inferior extensor retinaculum

Insertion Into base of proximal phalanx of great toe, into lateral sides of tendons of extensor digitorum longus of second, third, and fourth toes

Action Extends the four toes

Nerve Deep peroneal nerve (L5, S1)

ABDUCTOR HALLUCIS

(First layer)

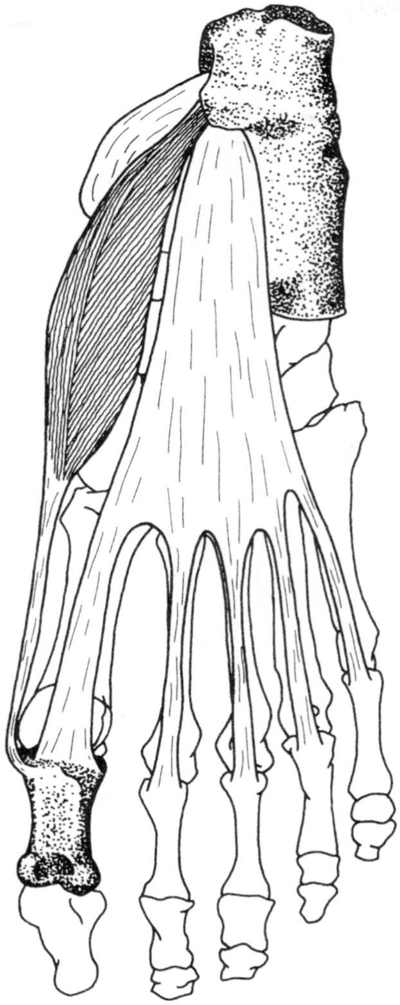

Foot—plantar view

Origin Tuberosity of calcaneus, flexor retinaculum, plantar aponeurosis

Insertion Medial side of base of proximal phalanx of great toe

Action Abducts great toe

Nerve Medial plantar nerve (L4, L5)

Note: The muscles of the sole of the foot can be divided into four layers (from superficial to deep):

First layer—abductor hallucis, flexor digitorum brevis, abductor digiti minimi

Second layer—quadratus plantae, lumbricales (tendons of flexor hallucis longus and flexor digitorum longus pass through this layer)

Third layer—flexor hallucis brevis, adductor hallucis, flexor digiti minimi brevis

Fourth layer—interossei (tendons of tibialis posterior and peroneus longus pass through this layer)

FLEXOR DIGITORUM BREVIS

(First layer)

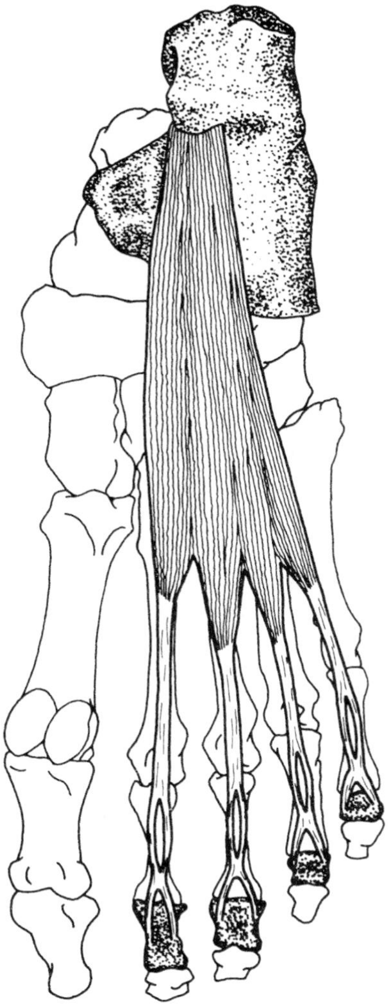

Foot—plantar view

Origin Tuberosity of calcaneus, plantar aponeurosis

Insertion Sides of middle phalanges of second to fifth toes

Action Flexes proximal phalanges and extends distal phalanges of second through fifth toes

Nerve Medial plantar nerve (L4, L5)

ABDUCTOR DIGITI MINIMI

(First layer)

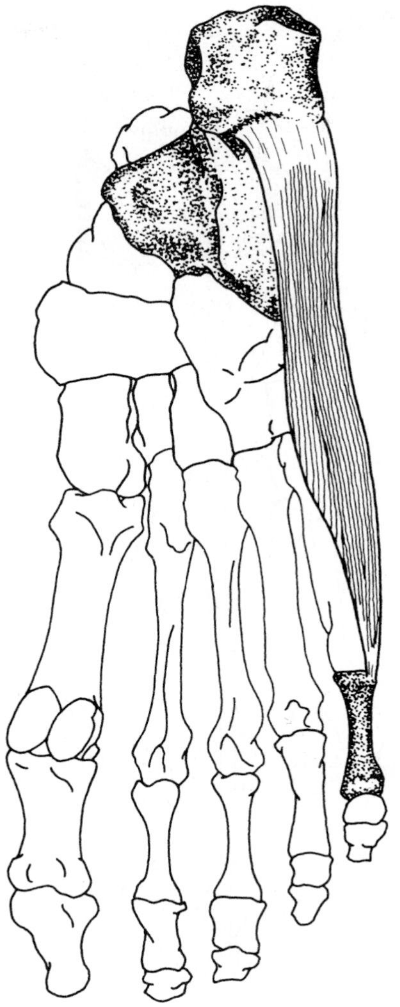

Foot—plantar view

Origin Tuberosity of calcaneus, plantar aponeurosis

Insertion Lateral side of proximal phalanx of fifth toe

Action Abducts fifth toe

Nerve Lateral plantar nerve (S1, S2)

QUADRATUS PLANTAE

(Second layer)

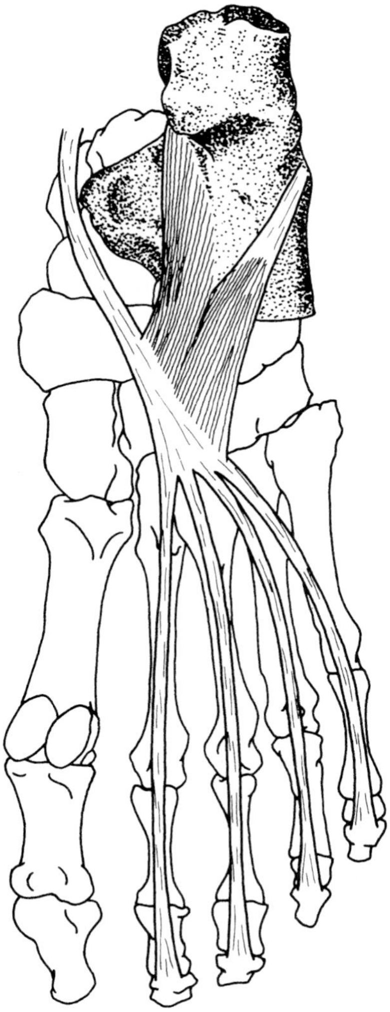

Foot—plantar view

Origin Medial head—medial surface of calcaneus

Lateral head—lateral border of inferior surface of calcaneus

Insertion Lateral margin of tendon of flexor digitorum longus

Action Flexes terminal phalanges of second through fifth toes

Nerve Lateral plantar nerve (S1, S2)

LUMBRICALES

(Second layer)

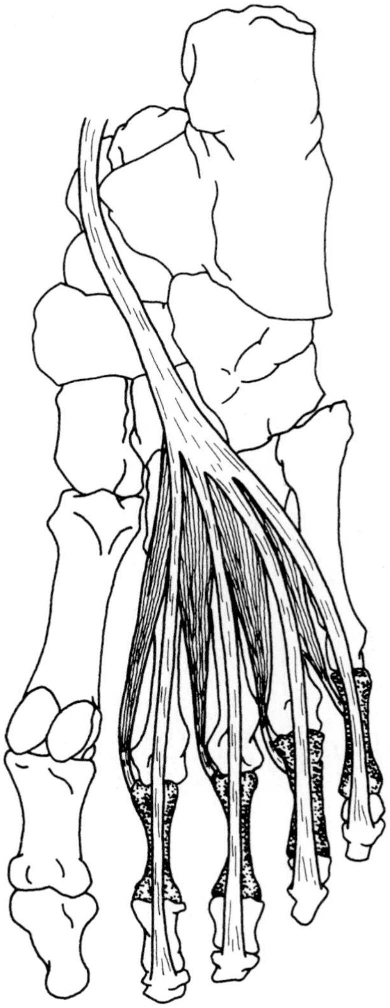

Foot—plantar view

Origin Tendons of flexor digitorum longus

Insertion Dorsal surfaces of proximal phalanges

Action Flex proximal phalanges of second through fifth toes

Nerve First lumbricalis—medial plantar nerve (L4, L5)

Second through fifth lumbricales—lateral plantar nerve (S1, S2)

FLEXOR HALLUCIS BREVIS

(Third layer)

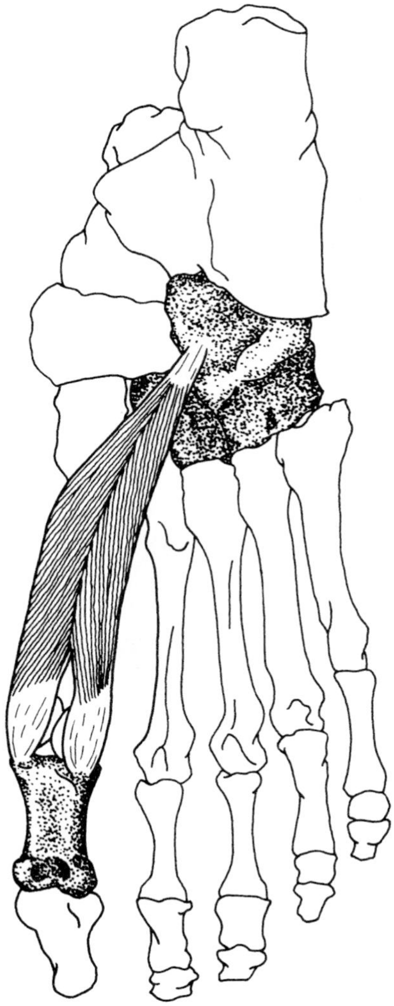

Foot—plantar view

Origin Cuboid bone, lateral cuneiform bone

Insertion Medial part—medial side of base of proximal phalanx of great toe

Lateral part—lateral side of base of proximal phalanx of great toe

Action Flexes proximal phalanx of great toe

Nerve Medial plantar nerve (L4, L5, S1)

ADDUCTOR HALLUCIS

(Third layer)

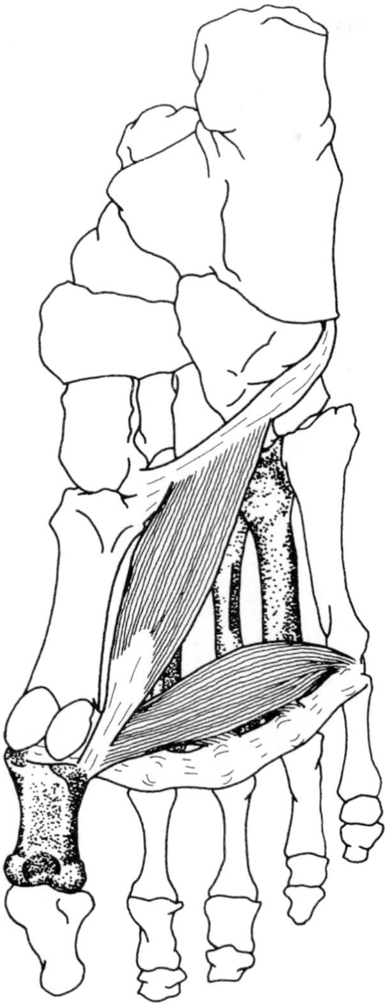

Foot—plantar view

Origin

Oblique head—second, third, and fourth metatarsal bones, and sheath of peroneus longus tendon

Transverse head—plantar metatarsophalangeal ligaments of third, fourth, and fifth toes, and transverse metatarsal ligaments

Insertion Lateral side of base of proximal phalanx of great toe

Action Adducts great toe

Nerve Lateral plantar nerve (S1, S2)

FLEXOR DIGITI MINIMI BREVIS

(Third layer)

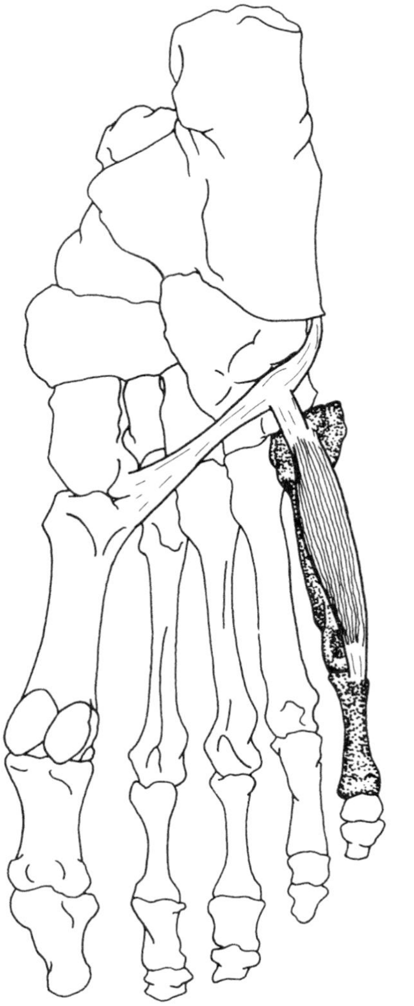

Foot—plantar view

Origin Base of fifth metatarsal, sheath of peroneus longus tendon

Insertion Lateral side of base of proximal phalanx of fifth toe

Action Flexes proximal phalanx of fifth toe

Nerve Lateral plantar nerve (S1, S2)

DORSAL INTEROSSEI

(Fourth layer; four muscles)

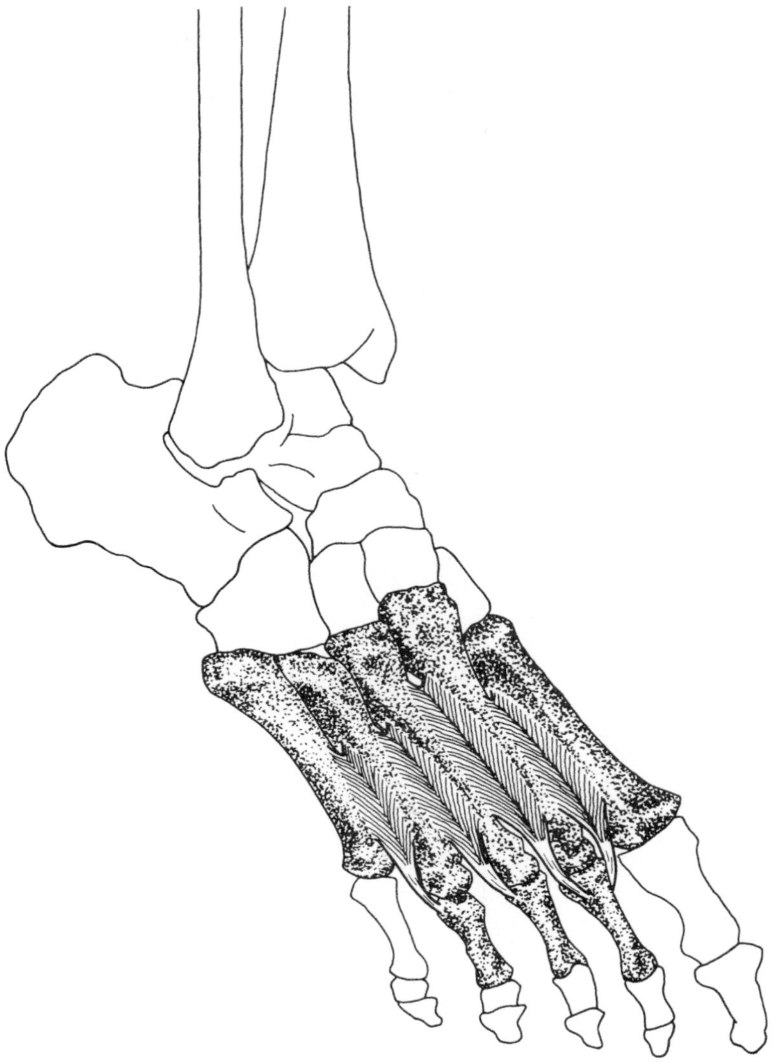

Foot—anterolateral view

Origin Adjacent sides of metatarsal bones

Insertion Bases of proximal phalanges
First—medial side of proximal phalanx of second toe
Second, third, fourth—lateral sides of proximal phalanges of second, third, and fourth toes

Action Abduct toes, flex proximal phalanges

Nerve Lateral plantar nerve (S1, S2)

PLANTAR INTEROSSEI

(Fourth layer; three muscles)

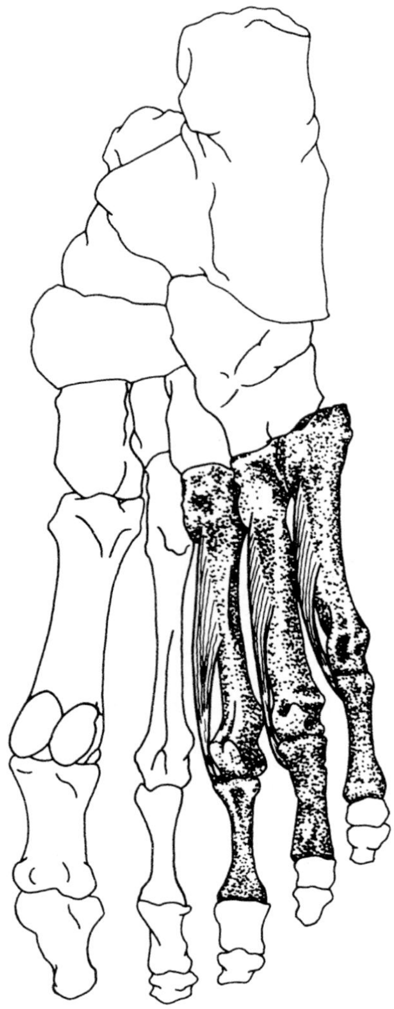

Foot—plantar view

Origin Bases and medial sides of third, fourth, and fifth metatarsal bones

Insertion Medial sides of bases of proximal phalanges of same toes

Action Adduct toes, flex proximal phalanges

Nerve Lateral plantar nerve (S1, S2)

Alphabetical Listing of Muscles

Index

T

U

V

Z

Notes

Notes

Notes

Notes

Notes

Notes

Notes

Notes

Notes

Notes

Notes

Notes

Notes